Basic Sys nalysis

Hutchinson Computer Studies Series

Basic Systems Analysis
Barry S. Lee

Basic Systems Design
Joseph E. Downs

Computer Appreciation
Allan D. McHattie and Jennifer M. Ward

Computer Systems: Software and architecture
John Newman

Computing in a Small Business
Howard Horner

Data Processing Methods
Barry S. Lee

Fundamentals of Computing
Neil A. Sheldon

Information Systems
Ian A. Beeson

Introduction to Program Design, An
Rod S. Burgess

Scientific Programming
William M. Turner

Management of Systems Development
Robert Bell

Structured Program Design Using JSP
Rod S. Burgess

Basic Systems Analysis

Second edition

Barry S. Lee

Head of School of Computing, Manchester Polytechnic

Hutchinson

London Sydney Auckland Johannesburg

Hutchinson Education

An imprint of Century Hutchinson Ltd,
Brookmount House, 62-65 Chandos Place, London WC2N 4NW

Century Hutchinson Australia (Pty) Ltd
89-91 Albion Street, Surry Hills
New South Wales 2010, Australia

Century Hutchinson New Zealand Limited
PO Box 40-086, Glenfield, Auckland 10, New Zealand

Century Hutchinson South Africa (Pty) Limited
PO Box 337, Bergvlei 2012, South Africa

First published in Input Two-Nine Ltd 1980
Second edition published by Hutchinson 1983
Reprinted 1986, 1987, 1989

Set in VIP Plantin by
D.P. Media Limited, Hitchin, Hertfordshire

Printed and bound in Great Britain by
Courier International Limited, Tiptree, Essex

British Library Cataloguing in Publication Data

Lee, Barry
Basic systems analysis — 2nd ed.
1. System analysis
I. Title
003 QA402

ISBN 0 09 154091 7

To David and Richard

Contents

Editor's note

This book is one of a series of textbooks with a modular structure aimed at students of computer studies and designed for use on courses at most levels of academic and professional qualification. A coherent approach to the development of courses in computing has emerged over the last few years with the introduction of the BTEC National, Higher National and Post-Experience Awards in Computer Studies. The syllabus guidelines for these courses have provided the focus for this series of books and this ensures that the books are relevant to a wide range of courses at intermediate level.

Many existing books on computing cause frustration to teachers and students because, in trying to be all embracing, they usually include irrelevant material and fail to tackle relevant material in adequate depth. The books in this series are specific in their treatment of topics and practical in their orientation. They provide a firm foundation in all the key areas of computer studies, which are seen as: computer technology; programming the computer; analysing and designing computer-based systems; and applications of the computer.

There are ten books in the series.

Computer Appreciation is the introductory book. It is intended to put the computer into context both for the layman who wants to understand a little more about computers and their usage, and for the student as a background for further study. *Computing in a Small Business* is aimed specifically at the small businessman, or at the student who will be working in a small business, and sets out to provide a practical guide to implementing computer-based systems in a small business. It is a comprehensive treatment of most aspects of computing.

Fundamentals of Computing looks in considerably more depth than the previous two books at the basic concepts of the technology. Its major emphasis is on hardware, with an introduction to system software. *Computer Systems: Software and architecture* develops from this base and concentrates on software, especially operating systems, language processors and data base management systems; it concludes with a section on networks.

An Introduction to Program Design is about how to design computer programs based on the Michael Jackson method. Examples of program code are given in BASIC, Pascal and COBOL, but this is not a book about a programming language since there are plenty of these books already available. This title complements *Program Development: Tools and techniques*, which looks at the task of programming from all angles and is independent of program design methods, programming languages and machines.

Data Processing Methods provides a fairly detailed treatment of the methods which lie behind computer-based systems in terms of modes of processing, input and output of data, storage of data, and security of systems. Several applications are described. *Information Systems* follows it up by looking at the role of data processing in organizations. This book deals with organizations and their information systems as systems, and with how information systems contribute to and affect the functioning of an organization.

Basic Systems Analysis offers an introduction to the knowledge and skills required by a systems analyst with rather more emphasis on feasibility, investigation, implementation and review than on design. *Basic Systems Design*, the related volume, tackles design in considerable depth and looks at current methods of structured systems design.

The books in this series stand alone, but all are related to each other so that duplication is avoided.

Barry S. Lee
Series editor

Preface

This book is aimed at giving an overview of the knowledge and skills required by a systems analyst. It assumes that the reader has a basic understanding of computer technology, programming and data processing methods, such as would be gained from reading other books in the series. It does not treat in detail the design aspects of the analyst's job because this topic is covered fully in the companion volume, *Basic Systems Design*.

The book begins by describing the nature of data processing systems in which the systems analyst would be interested (two other books in the series, *Data Processing Methods* and *Information Systems*, provide a much more detailed treatment). It then looks at the nature of the systems analyst's job and the stages through which the development of new computer-based systems goes. Each of these stages is then examined in a separate chapter: Feasibility study, Investigation, Design, Implementation, and System maintenance and review. The final chapters concentrate on the communication aspects of the systems analyst's job in terms of communication skills and the documenting of systems.

The book uses the NCC *Data Processing Documentation Standards* throughout, and there are practical exercises at the end of each chapter, to which suggested solutions are given at the back of the book.

Acknowledgement

The author gratefully acknowledges the permission granted by the National Computing Centre Ltd, Oxford Road, Manchester M1 7ED to use various forms and standards from their *Data Processing Documentation Standards Manual*.

1 Data processing systems

Data processing

Although the word system is used to describe a wide variety of phenomena which we encounter in our daily lives, the systems analyst is usually involved just with systems within organizations that are concerned with processing data. It is important that a book of this kind begins, therefore, by defining what we mean by data, processing and systems before going on to examine the role of the systems analyst.

Data

Data can be defined as a set of symbols which are used to represent objects, events or activities within the real world as we know it. For example, in conversation about the objects that we see in our dining room at home, we use words like table, chair, carpet, door, light etc. in order to convey ideas to other people. We do this on the assumption that we have a common understanding of what a 'door' or a 'table' or a 'chair' is and we use the words as a shorthand way of defining the object that we are discussing. In other words 'door' is a code for the object to which we are referring; the code could just as well have been 18753/21, so long as we all know that that code refers to a door. Similarly we can refer to people by their name, e.g. Barry, or we can give them a number such as a student number. In computer-based systems where precision in identifying objects is required, we usually find that numbers are used to identify them partly because numbers can be easily represented within the computer and partly because they can be readily manipulated.

Just as objects can be coded, so can events and activities. For example, when you make a purchase in a department store, that activity can be coded by its date, the amount involved, the code number of the item purchased, and the type of transaction (i.e. whether you paid by cash or by credit card). Or again, events (such as each time a car passes a sensing point) can be recorded in coded form for subsequent processing.

Processing

Normally data can be seen as a collection of symbols, meaningless until they are processed. The purchase in the department store mentioned above might be recorded as 19047900126913876547 which is just a string of digits until use is made of it. The processing of data involves the execution of various operations on the data (i.e. manipulation of the data) into a form which is meaningful to a human being or another machine. When it is meaningful, it is usually described as information.

Processing can include arithmetic operations (add, subtract, multiply, divide), logic operations (checking, testing, comparing) or simple movement operations (editing, transmitting, displaying, recording). So the string of digits above might be processed in a variety of ways:

1. It might be merely edited into its separate elements, viz.
   Date of purchase 19/04/79
   Amount of purchase £12.69
   Code number of item 1387654
   Code of transaction type 7 (= cheque).
2. It might be sent along a telephone line and then displayed on a television screen.
3. It might be stored with all the other purchases for that day and then analysed to see how many items with code number 1387654 were sold, or what was the total value of purchases, or how many customers paid by cheque.

and so on. Once the string of characters is processed in one of these ways, the result is *information*. Of course, it is usually the case that information at one level of operation is data at the next level.

For example, a student is interested in how many O level examinations he is going to sit individually – and that is information to him. The examining board uses the data about individual examination entries to compile information about the total number of students for different examinations for which it has to print examination papers. So in this instance what is information for the student is data for the examining board.

Thus data processing can be considered to be the manipulation of symbols to produce information which is useful to the recipient.

Systems

A system is usually defined as a set of interrelating elements which come together to achieve a specific objective. We have a central heating system consisting of a boiler, hot water, pipes, radiators and thermostats which operate together in order to heat a house. There are electoral systems which govern the way people vote and their votes are counted in order to elect a government. A clock can be considered as a system made up of balances, wheels, fingers, numbers etc. whose purpose is to show the time; and so on. Almost everything we deal with in life can be described in terms of a system.

In the case of data processing systems, we are concerned with the elements which interrelate in a systematic way to process data into information; these elements will consist of procedures, rules, files, computers, pens, paper, and, above all, human beings. A typical data processing system in an organization is the payroll system; in simple terms it collects data about hours worked, relates it to rates of pay and calculates wages; humans are involved at various stages of the processing and often machines are involved also. It is the way in which this interaction occurs that it is of interest to the systems analyst whose job is to analyse the way things happen at present and to see whether improvements can be made by reorganizing the procedures or introducing new equipment.

We will now go on to look at some more advanced ideas about systems and data processing.

Systems

As soon as one moves away from purely physical systems like the central heating system to systems which involve human beings, they become very complex to study and so we need some guidelines to help us to analyse the nature of the system. These guidelines are provided by systems theory.

System identification

First of all we need to be able to identify a system. This is extremely difficult because all systems are part of other systems (i.e. subsystems of other systems) and contain within themselves subsystems. For example, at any point in time you are part of a number of social systems (e.g. the educational system, the political system, the school or college system – you are a subsystem within each of these); but there are subsystems at work within you (e.g. circulatory system, respiratory system – they are subsystems within you).

Systems are usually identified by the elements of which they are composed and these elements delineate the boundary of the system. Outside the boundary is the system's environment which usually affects the way a system operates. Relationships between separate systems are called

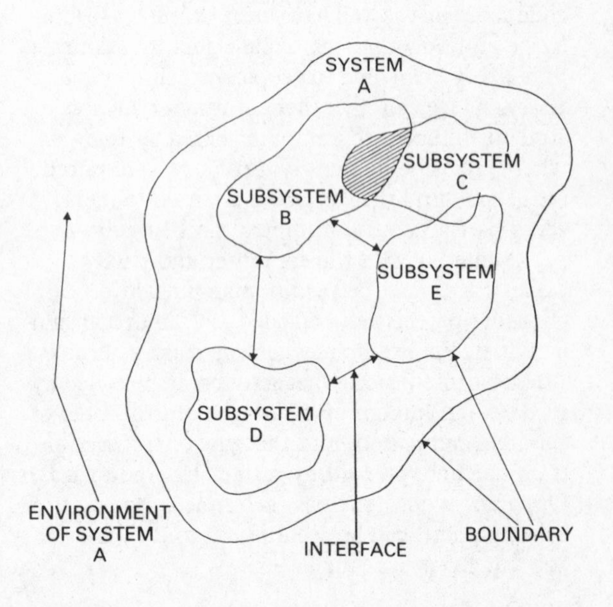

Figure 1 *System, subsystem, boundary, environment, interface*

interfaces and each subsystem may well interface with every other subsystem within a system. The interface normally takes the form of the passage of data or information between the subsystems. Diagrammatically these concepts can be represented as Figure 1 (note that occasionally subsystems overlap, as with subsystem C and subsystem B on the diagram; the shaded portion is common to each subsystem).

Using these concepts, we can identify organizations as systems. If we describe a business as a system, then it will have subsystems concerned with marketing, production, accounting etc. The subsystems of any particular organization will be determined by the objectives of that organization. Each of the subsystems will have a boundary (identified by its elements such as people, offices, equipment, files, jobs); and there will be interfaces between the subsystems in the form of information which passes between them – for example, the marketing subsystem will tell the production subsystem how much it intends to sell, the production subsystem will tell the marketing subsystem how much it can produce, and the accounting subsystem will tell each of the other two how much cash is available to meet their requirement. Clearly each subsystem is influenced by the other subsystems which form its environment; and the system as a whole is influenced by its environment (e.g. customers, suppliers, government, general public etc.).

System definition

Having identified the systems, we next need some way of defining them. This is done using a simple model of a system as shown in Figure 2. The system is seen as a set of processes which receive inputs and produce outputs. The nature of the inputs, processes and outputs is governed by the objectives which the system is aiming to fulfil. Thus a typical manufacturing system receives raw material and converts it by a series of processes into products which can be sold; and a typical data processing system receives input data such as hours worked per employee and converts them into information such as a payslip which enables the employee to be paid.

The trouble with this simple model, however, is that it does not show any means of checking whether the system is performing adequately and so an extra element has to be introduced, known as control – as shown in Figure 3.

Here a sample of the outputs is fed to a measurement or control function and compared to the output standard required by the system objective. If, as a result of measurement and control, something is felt to be in need of adjustment, then feedback can be provided to the inputs/processes to bring about the adjustment. For instance in the manufacturing system, if the output products are not being produced at a sufficiently high quality, then the raw material or production processes can be reviewed and improved. The control mechanism allows the system to regulate itself in an effort to achieve the objectives in a continuing way.

Once this control facility is built in to systems we can next turn to the relationship between a system and its environment. If a system was left to continue its processing regardless of the environment, then eventually it would decline (this is known as entropy). For example, if a business continues to manufacture nylon stockings for women when demand has changed to tights, it will soon go out of business. To survive an organization has got to be sensitive to its market environment and has to adjust to changes in that environment. This brings us to the concept of an adaptive system in which information about the environment is fed to the system to allow it to adjust its input, processes and even its objectives. This ability to react to the environment tends to make systems more stable.

Building an approach

Having identified and defined systems, the final requirement is build an approach to studying systems. Normally the approach is one of taking a total system and breaking it down into its subsystems – the process is known as factoring. The total system will vary from study to study. In one study the total system may be the business; in another it may be the marketing subsystem of the business; in another it may be the sales order processing subsystem within marketing. The approach is to take the total system and to factor it

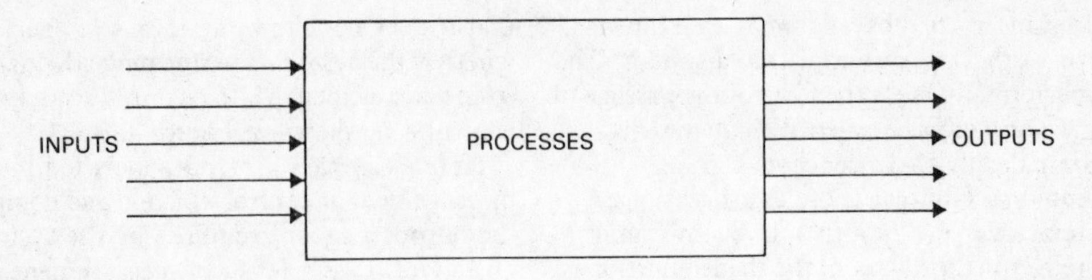

Figure 2 *A system model*

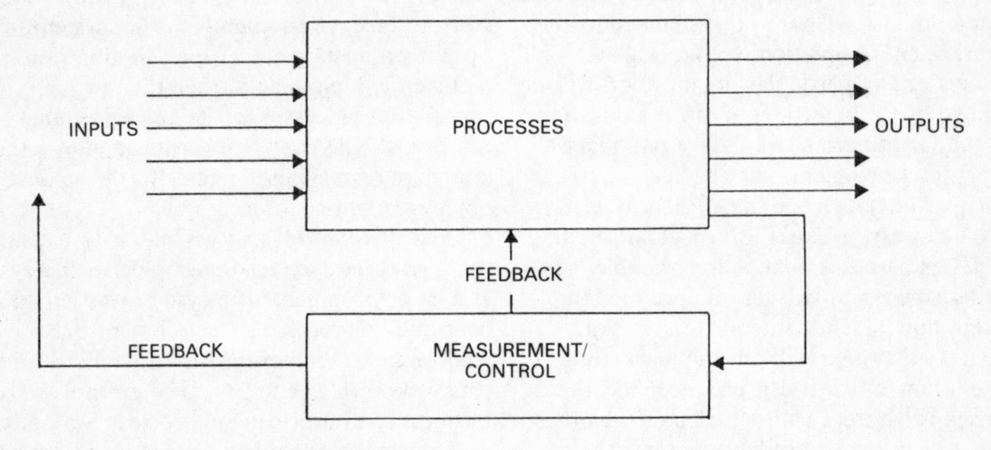

Figure 3 *A control system model*

into a series of subsystems; to analyse and improve these subsystems; and then to integrate them. Integration is a fundamental feature of systems theory. It draws attention to the primary importance of the total system which of course determines the role of the subsystems. Subsystems have no existence except as part of the whole system and so they must be directly related to it. The difficulty of course lies in the identification of the whole system – and this will usually be done arbitrarily dependent on the political circumstances surrounding the system study. Thus in examining the marketing subsystem of an organization the emphasis must always be on its relationship to the organization as a whole; it is this which will determine the objectives of the marketing subsystem and shape the processes of its subsystems.

Information flow

It will be evident from the discussion of systems that every system is dependent on data. The interfaces between the subsystems are lines of communication along which data is passed to regulate the operation of the subsystem. Information from the environment of the subsystem is crucial if it is to be stable. Information about performance is required to provide feedback to inputs and processes in order that they can be adjusted.

Every organization carries out data processing activity in order to control its subsystems. In the one-man business the data required is kept in the head of the proprietor (or perhaps on scraps of paper). As the business expands, more people are employed, specialists are used, more processes are carried out, greater stocks are held, and so on; as a result more information is needed, and the data gathering, storing, manipulation and retrieval needs to be formalized. This is where data processing systems come into play, keeping track of the activities of the subsystems.

For example, before the production subsystem can start to make products, it needs to know what type, size, colour, packaging etc., how many to produce, at what time and in what location, how much raw material to purchase, how far in advance and how it is to be stored. One could go on almost endlessly. This information must be provided by data processing. Market research, orders, supplier performance, current stock levels, prices of raw material are all elements of data which need to be analysed to answer the questions.

The job of the systems analyst is to identify the information required by the various subsystems to allow decisions to be made and to define a data processing system to produce such information. As was said earlier, all organizations have data processing systems performing this function already but sometimes they do not work very well or are not sufficiently comprehensive or are distorted by department boundaries. The systems analyst must examine what exists already, correct/remove those parts which are inefficient, retain those parts which work perfectly well and introduce new procedures to fill the gaps where they exist. This ultimately involves designing a new data processing system to provide the required information. Very often computers will be involved.

The computer

The computer is an ideal tool for processing data but it should only form part of the systems analyst's solution if it proves to be the most economical tool. The advantages it offers are:

1 Speed of access and manipulation of data;
2 Accuracy and reliability in processing data;
3 Storage and retrieval of large volumes of data;
4 Accessibility of data;
5 Central storage of data.

The disadvantages are less clear. Some would argue that it brings about too much standardization, and tends to make jobs much less challenging; others would argue quite the opposite.

The activities of data processing identified earlier – collection, storage, manipulation and retrieval of data – are usually greatly improved by computer usage for any system other than a very small one. Data can be captured by very sophisticated methods often directly into the computer; it can be stored in a very small area; it can be manipulated at high speed; and it can be made available in a variety of forms (provided these have been foreseen) at almost the flick of a button. Thus the computer is bound to loom large in the range of tools available to the systems analyst for data processing systems.

An example

We will finish this chapter by taking an example of a supermarket, in a simplified form.

If we examine the supermarket operation in systems terms we will find that it has a large number of subsystems. Using the simple model of a system, we can say that the supermarket's inputs

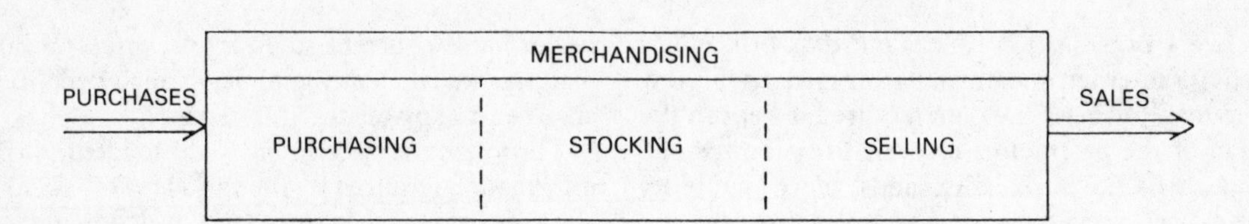

Figure 4 *A simple system model of a supermarket*

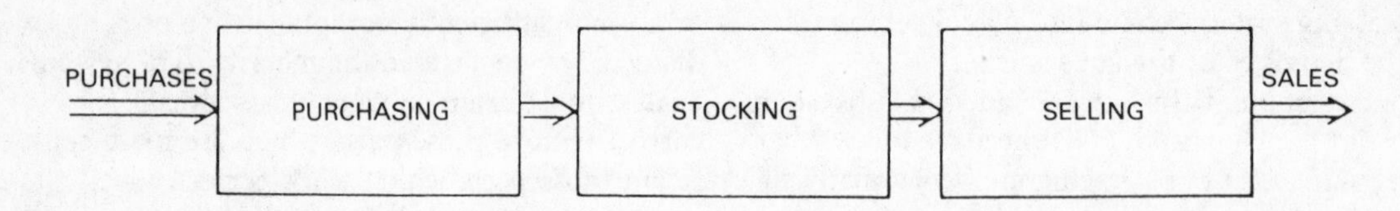

Figure 5 *An alternative model*

are purchased goods and its outputs are sales by retail to customers. The three most obvious processes then are purchasing, stocking and selling (all of which are often called merchandising) which could be represented diagramatically as Figure 4. Normally as the merchandising operation increases in size, so there is a tendency for specialization and factoring of the overall system into a number of subsystems. Thus we would probably find the three processes identified above as separate subsystems, as in Figure 5.

Each of the processes needs to be managed (or planned and controlled) and this introduces the concept of a control loop for each of the processes – and of course a control loop for the organization as a whole. Thus our system model begins to look more like Figure 6. In this diagram the physical flow of goods is represented by the double-line arrow – all the single-line arrows represent the flow of information necessary to control the physical flow of goods to keep it in line with overall company aims. The nature of this information flow in simple terms is as follows (the numbers relate to Figure 6).

The management of the organization (i.e. overall control) is provided (by the board of directors/shareholders in a private enterprise organization) with a statement of objectives and standards of performance (1).

These are interpreted by the overall control subsystem and passed on to the control subsystems for each of the functions as a set of budgets and operational plans (2, 3, 4).

The operational plans are then put into action. Instructions are given for contracts to be negotiated and goods to be purchased in appropriate quantities (5); for stocks to be held and displayed and checked to achieve an adequate rate of turnover (6); and for goods to be advertised and promoted to encourage sales (7).

Feedback is provided to the control subsystems from each of the operational units so that performance can be measured, e.g. information on quality of goods purchased, delivery times, prices etc. from purchasing (8), rates of stock turnover, deterioration of goods, pilfering etc. from stocking (9), and sales quantities, sales value, types of customer, product preferences etc. from selling (10). This feedback enables adjustments to be made to the way the processes are conducted (5, 6, 7) or to the volumes/types of input to each of the processes (11, 12, 13).

At the same time the overall control subsystem receives feedback in summarized form from the operational units (14, 15, 16) and from the control subsystems (17, 18, 19). As a result of this the budgets and operational plans may be adjusted (2, 3, 4) or feedback may be provided to the board of

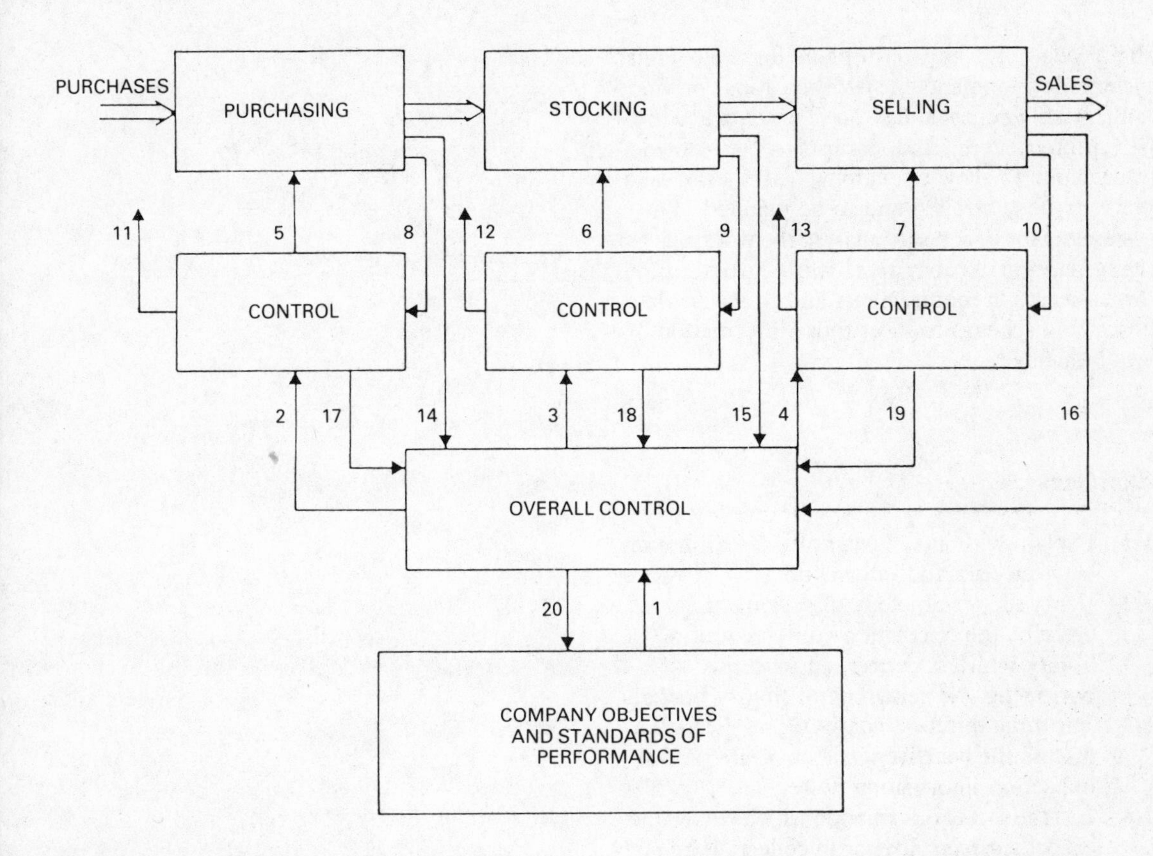

Figure 6 *A simple control system model for a supermarket*

directors to cause a change to the company objectives/standards of performance (20), which is then interpreted by the overall control function and the cycle begins again.

Of course, our model is greatly simplified. We have not introduced the many other supporting subsystems of a supermarket which may be concerned for example with transportation of goods to the supermarket, accounting for cash flows (for payments of suppliers, employees etc. and receipts for sales from customers), recruitment and training of staff, buildings and equipment maintenance and protection etc. These would each need to be controlled and interrelated and would each have significant information requirements. For instance, each item purchased would need to be recorded in order to arrange payment for it, and to make the stocking subsystem aware of its availability; data about each employee is needed in order to schedule jobs, organize payments, deduct tax and national insurance contributions etc. So even a simple system begins to be very complex.

The requirement for data processing within the supermarket system is clear. There is need for data about goods purchased and available for purchase (such as type, price, supplier, quantities etc.); about goods in stock (such as quantity, condition, length in stock, value etc.); about goods sold (such as quantity, value, customer, VAT rate etc.); about employees (personal details, working hours, rates of pay, tax codes etc.); about equipment and facilities (selling space, shelves sizes, vehicles, cash registers etc.); about cash flows (cash coming in through sales, cash going out for purchases); about markets (type of customers, money available, preferences, use of advertising/promotional ideas);

about long- and short-term plans for the overall system (developments in space, location, product range, staffing etc.); and so on. This data needs to be captured, stored and manipulated to provide information to allow operational activities to take place, to be controlled and to be planned. The systems analyst's job is to analyse the workings of the system and its subsystem in order to determine the information requirements and to devise data processing systems to meet those information requirements.

Exercises

1.1 Explain with lots of examples the difference between data and information.

1.2 Using the terminology of systems theory, describe the circulation subsystem of a library which is concerned with the borrowing and returning of library books.

1.3 For three different computer applications, discuss the benefits of the computer and its impact on information flows.

1.4 List the sort of information which you would expect a careers adviser to collect, store and retrieve. Suggest ways in which a computer could help.

2 Systems analysis

The job of systems analysis

The job of the systems analyst is to investigate and assimilate information about the way a system currently operates; to analyse its performance in the light of the system objectives which are identified by management; to develop and evaluate ideas about how the system can be improved/reorganized; to design in detail a new system meeting the requirements which have been identified; and to implement the new system once it has been fully developed. The aim is to produce a new system which will operate effectively, efficiently and economically; this may or may not involve the use of a computer (although most systems which the systems analyst designs will be computer based, the primary aim is not 'to computerize'). The systems with which he deals are data processing systems which are open, dynamic and probabilistic, involving human beings whose needs have to be taken into account throughout the investigation, analysis, design and implementation processes. The design effort is directed not towards theoretical excellence but rather towards practical and flexible solutions to real-life problems in industry and commerce.

The task of systems analysis is not a new one; it has been carried out for a long time under a variety of names. Work study engineers, organization and methods (O & M) officers, operational researchers (OR) have been (and still are) concerned with very much the same problems as the systems analyst, though usually from a narrower viewpoint. Work study examines work methods and techniques on the shop floor; O & M concentrates on clerical methods and procedures; and OR looks for quantitative solutions to business problems. Systems analysis in a sense encompasses the other three and in addition is concerned with the possibility of computer assistance in solving the problems of information flow. The very nature of systems analysis is to examine the total system in its environment rather than parts of it.

Nowadays the job title 'systems analyst' can cover a variety of different jobs, and, as the need for systems analysts has grown (with increasing use of computers), so there has been a tendency for specialization within the broad area of systems work. For example, the person whose work concentrates mainly on investigating existing procedures and identifying user requirements (without going as far as designing a new computer-based system) tends to be known as a 'business analyst' or 'information analyst'. The person who takes the statement of user requirements and turns it into a specification of a new (computer-based) system will be known as a 'system designer' or 'computer system designer'. The person whose main effort is devoted to the design of files and data structures across a range of systems may be known as a 'file analyst' or 'database analyst'. And the person who is employed to carry out the implementation of systems is often called a 'system implementer'. All of these jobs may be subsumed under the title 'systems analyst' and so in this book we examine the total job of the systems analyst which includes all the specialisms identified above.

Role of the systems analyst

In carrying out these tasks the systems analyst essentially acts as a link between users of the computer and the computer itself in order that the users can improve their information flow by making use of the computer. This linking role has a variety of aspects and perhaps we can look at them briefly under the following headings.

Catalyst

Above all else, the systems analyst acts as a catalyst providing an opportunity for users to examine closely their job function and their use of information in order to determine how a new system might most adequately serve them. By challenging people to think objectively and perceptively about their current situation and their information needs, the systems analyst forces them to question what they are doing and why they are doing it – and this should give the user a better understanding of his own job.

Adviser

The systems analyst is an adviser to the user about what can and what cannot be done by using the computer. He should never take an executive role, deciding on the nature of a system – all decisions of this kind lie in the hands of the managers of the department which will use the system with guidance for the systems analysts.

Educator

On the whole the systems analyst will be dealing with users who have little or no knowledge of computers or of his job function. Thus in every situation of contact with users, both formal and informal, the systems analyst has a responsibility for educating staff and for encouraging participation in investigation and design. Help and support needs to be given to those who are afraid of or do not understand the technology. Commitment of users to a new system will not be achieved by the systems analyst going off and designing a system which he finds acceptable; the users must be involved in identifying their own information

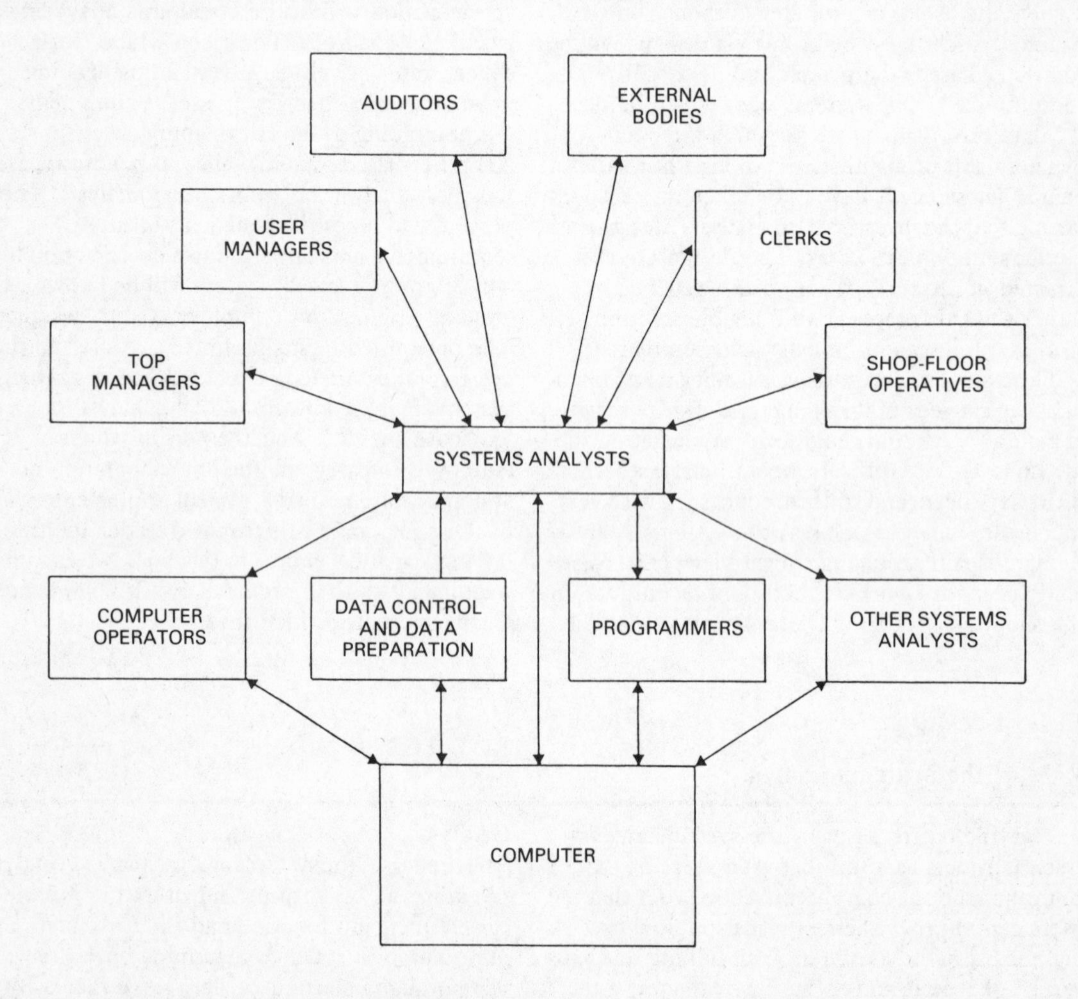

Figure 7 *The systems analyst's communication links*

needs and controlling the design of a system to meet their requirements. The systems analyst's role is to contribute technical expertise and systems methodology to the exercise.

Salesman
At many stages in a system development project, the systems analyst will need to sell his ideas to the managers who decide whether a project should go ahead, and to try to convince users at all levels that a new system will be an improvement and not a hindrance to job satisfaction and performance.

Communicator
Communication skills are essential to the systems analyst. He needs to put ideas across to and interpret the requirements of a wide range of different people (Figure 7). He is involved with the users of the system, both within and outside the organization, with external bodies such as government agencies and auditors, with technical staff such as computer programmers and operational researchers, with computer operations staff such as operators, data control clerks and data preparation clerks, and with other systems analysts working on allied projects. He needs to communicate at all levels of the organization with people from very varied social backgrounds and educational levels.

Agent of change

The most frequently discussed role of the systems analyst is that of an 'agent of change'. The systems analyst is seen as a person who brings, by virtue of new technology and investigation of current methods, change to people's job situation. People on the whole are not very keen on change, especially changes which are imposed on them by external pressures, and so the systems analyst has to learn to handle the introduction of change carefully. This is not easy because the computer is surrounded by fear and suspicion even after 20 years of usage.

The public image of the computer is poor. On most occasions, when it is discussed in the press or on television, the approach is critical or far-fetched – the computer is highlighted for the mistakes that are made (even though these are rare and always caused by human error) or for its superhuman powers. As a result people have a very odd impression of the machine.

In addition the systems analyst is often seen in a hostile light as a well-paid technocrat whose interest lies in a satisfyingly designed system at the expense of those whose lives and livelihood are affected by the design. The high salaries of computer staff and their loyalty to the technology rather than the organization, as seen by the users, tends to separate the systems analyst from the people whom he should be trying to help. The users are often afraid that computer systems may lead to loss of jobs or retraining or redeployment; they are suspicious of management's motives in seeking to introduce computer systems; and they resent the implicit criticism of a systems analyst investigating their job. Generally these fears are misplaced; for example, there is plenty of evidence that computer systems do not lead to wide-scale redundancy. But lots of factors within the organization can cause such fears and doubts. It is very easy for false impressions to be gained through rumours about computer usage, and steps need to be taken to correct such false ideas.

Hostility to the computer manifests itself in a variety of ways to the systems analyst. He may find some people who are very aggressive towards him; others will just avoid him; others will quite innocently tell him variations of the truth which they feel puts them in a more favourable light; and others will find it very difficult to be helpful in any way. This kind of behaviour is not unusual in a situation of change and the systems analyst has to learn to cope with it. He has to accept that the social acceptability of a system is of paramount importance – often more important than technical excellence. He has to learn to treat each situation as unique and to use his social skills to achieve a rapport with user staff; each individual is unique and has his own ideas and must be treated as such by the analyst.

It is a plain fact that ideas cannot be successfully imposed on people; effort has to be placed on winning people round and persuading them of the benefits of any change.

There are many things that can be done to facilitate change in the user environment. Perhaps the most important is to ensure that people are kept well informed about everything that is going on – in the most open way possible – so that hostility cannot arise owing to lack of awareness. Allied to this is the importance of educating users about the nature of computer systems, their potential and their shortcomings. The more the users know, the better their contribution to the change is likely to be. This is particularly relevant where the systems analyst encourages a participative approach to system development and acts as an adviser to a group of users to enable them to design and develop their own new system. Besides information, education and participation, the systems analyst needs to deal sympathetically with users both as individuals and in social groups. He has to be seen to share at least an understanding of their goals and to deal honestly with them. This is difficult to achieve, particularly if the management style is autocratic, and the analyst should not be identified too closely with management's aims. He needs to be aware of and to emphasize the help available to individuals and to take advantage of formal consultation/negotiation mechanisms through trade unions, staff associations etc.

All of these activities are aimed at enabling users to be involved in the development of new systems. Without such involvement, the users will have great difficulty in accepting any new system and it will be misused or not used at all as a result. The analyst has both a moral and a professional duty to provide users with the occasion and if possible the power to influence the design of systems which affect their lives and livelihood. In the long run such involvement in any case will prove financially beneficial to the organization because of the greater user commitment to the new system.

Personal qualities required in a systems analyst

It will be clear from what has been said above that the systems analyst needs a range of personal qualities to achieve success. On the one hand he is dealing with people and has to win their trust; on the other he has to separate himself from those people in order to develop ideas in an objective way about the system. In other words he needs to combine the attributes that are normally associated with the extrovert and the introvert. In dealing with people, he needs above all to be acceptable; he may interview the managing director one day and the shop-floor operative the next, and so he has to have a range of social skills to suit people of widely varying backgrounds. He needs to be patient with the slow, perceptive with the quick; he must be confident so as to imbue confidence; he must be fair and unbiased in all his dealings; he must be persistent without being overbearing; he must be a good listener who knows when to interject. Perhaps the most appropriate word is empathy – he needs to be able to understand the problems and aspirations of those with whom he deals.

In analyzing and designing systems the major requirements are, to coin a phrase, 90 per cent perspiration and 10 per cent inspiration. The systems analyst needs to be patient and perceptive in objectively analysing events, going over them again and again in a painstaking way to ensure that the picture he has formed of the existing system or the design he has made for the new system is accurate. Care with documentation, analysis of data, accuracy of transmission of information etc. are crucial. All the creative genius in system design is of no use if it is not accurately and comprehensively communicated to those who have to implement or operate the system, and the latter involves hard, often monotonous, attention to detail.

Such characteristics are rarely found in one person!

Skills required by a systems analyst

The skills required by a systems analyst are almost as demanding as the personal qualities.

The first of these is the ability to recognize, define, describe and analyse a problem situation. Usually the analyst is working with undefined problems (in other words, the manager feels that he needs something to be improved but isn't quite sure what) and has to be able to 'home in' on the correct problem and its causes. Normally he will do this by presenting an objective, analytical and, above all, different view of the situation to the manager which will help to clarify what is required.

Having identified the problem the next skill is to develop a number of alternative approaches to 'solving' the problem in the recognition that no one solution ever exists. The evaluation of possible alternatives requires the application of knowledge from a wide range of disciplines, and judgement of the most appropriate ideas.

Third, these activities (and especially the design activity) have to be carried out in real life under pressures of time, money, politics, personality and technological change. The theoretical ideal which draws together several different perceptions has to be made to work in real life.

The fourth area of skill which is highly relevant to all of the previous three is communication skill. The analyst has to present his ideas and arguments clearly both in writing and orally; he has to extract and understand the arguments of others; he has to control interviews and meetings; and he has to sell his ideas to others. He needs to be adept at manipulating both words and numbers.

Finally he must be well organized. He will often find himself controlling a lengthy project with high expenditure on people's time and equipment and implications for other areas of the organization; realistic plans and careful control are required. The systems analyst therefore needs to be disciplined in his project approach, and reasonably flexible to cope with the inevitable difficulties which will be encountered along the way.

Knowledge required by a systems analyst

A full list of areas of knowledge which a systems analyst might require would be quite extensive, but would begin with three basic elements – computer technology and data processing techniques, concepts and techniques of systems analysis, and business background.

The systems analyst needs to be aware of the equipment and techniques available in using computers; this includes up-to-date knowledge of hardware, software and packages and some understanding of programming in order to achieve a smooth interface with programming staff. He needs to have a broad understanding of how business operates, the nature and flow of information in a business and the aims and processes of management. (Clearly academic knowledge of business is no substitute for experience.) To add to this knowledge of potential areas of computer application and potential methods of computer usage, the analyst needs a working knowledge of the concepts and techniques of systems investigation, analysis, design, implementation and evaluation.

Additional useful areas of knowledge would be organization and methods practice (to assist with the design of user procedures); operational research (so that OR can be applied to new systems); information and communication theory and skills; organization theory and practice; industrial/organizational sociology/psychology; ergonomics; quantitative techniques; and basic economics and accountancy. All of these subjects are relevant to particular aspects of the design activity.

Recruitment of systems analysts

Systems analysts tend to be recruited to their first appointment from one of three sources:

1 Those already employed in a related work area such as O & M, work study, OR or programming;
2 Those in user departments which have been or are likely to be affected by computerization; and
3 Graduates from degrees in computer studies (with a strong business orientation) or business studies (with strong computing orientation).

The most common route into systems analysis is from the first of these sources by promotion from programming. Most computer departments prefer their systems analysts to have had some programming experience – although it doesn't always follow that a person with programming ability will have the personality appropriate to systems analysis, which is usually more extrovert.

Recruitment procedures will invariably involve an aptitude test (normally used for negative screening) and a series of interviews in which the candidate will be required to demonstrate adequate knowledge and appropriate communication skills.

The data processing department

The department in which the systems analyst will work will usually be known as a data processing (DP) department or management services department.

A DP department will have a structure similar to Figure 8.

The data processing manager is responsible for planning and day-to-day control of the activities of the department. Normally he will divide responsibilities between a manager concerned mainly with development of new systems for the computer and a manager concerned mainly with running existing systems on the computer.

The system development manager will have both systems analysts and programmers reporting to him, often organized into project teams for

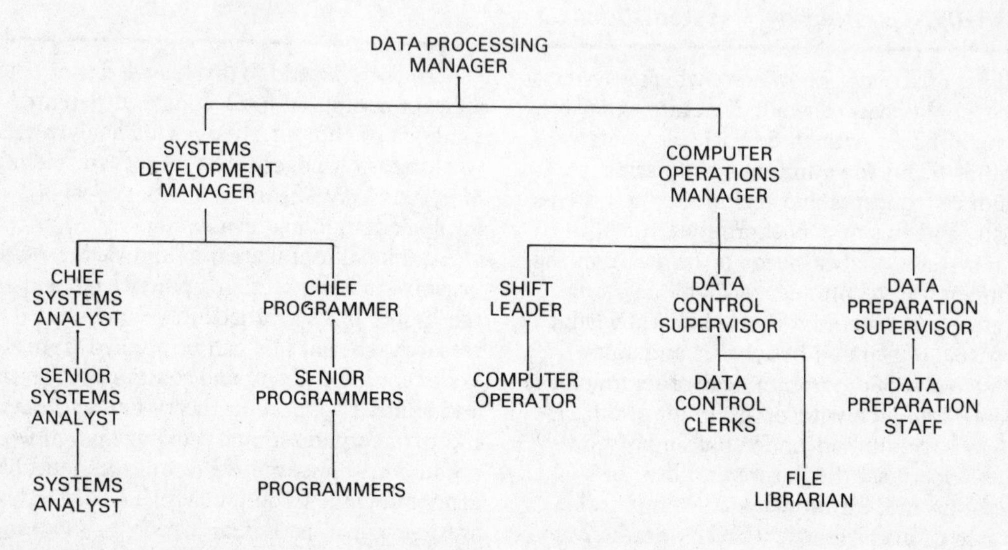

Figure 8 *DP department organization structure*

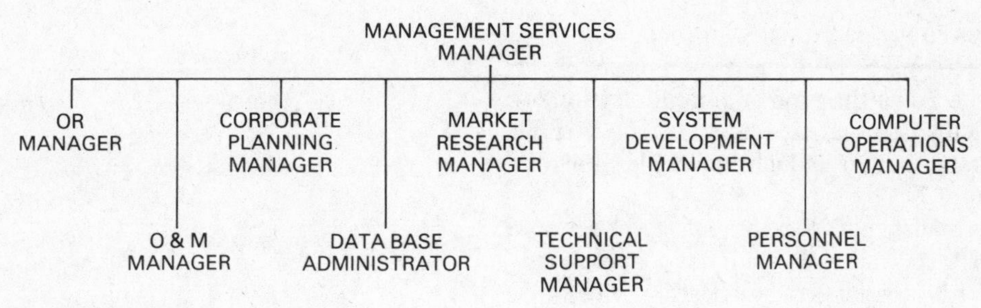

Figure 9 *Management services department*

particular exercises. Both analysts and programmers will be members of such teams. The analyst's job is to investigate, analyse, design and specify new systems – which the programmer will convert into instructions to be executed by the machine. Once the programs have been written, tested and documented, they are passed on to computer operations to be run as required. A considerable proportion of programming activity will be used to maintain such programs, and maintenance programmers may well be employed by the computer operations manager.

The computer operations manager is responsible for day-to-day running of existing computer-based systems. Thus he will have reporting to him data preparation staff (who convert input data into machine readable form), data control staff (who check the movement of data to and from the computer room) and computer operators (who run the jobs on the computer and sort out any error conditions or machine breakdowns).

Larger organizations have developed along the lines of a management services department as in Figure 9. In this structure, all the service activities which line management requires are brought together in one department concerned with information provision and planning. Thus many of the tasks of the systems analyst may be handled by specialists in O & M or OR or database administration.

Another development has been to remove computer operations out of the hands of a DP manager and into an office services department (on a par with the typing pool etc.). This places emphasis on the day-to-day service activity of the computer operation and is intended to avoid conflicts of priority which might arise if the operational activities are controlled by the person who is responsible for developing new systems.

Concluding comments

There is some debate as to whether systems analysis is a science or an art. Those who argue that it is a science point out that it is empirical, based on observation and measurement; it is based on logical analysis of quantitative information; its approach is methodical and systematic resulting in exact and factual conclusions; and it is objective and neutral in its impact on people.

On the other hand, those who believe that it is an art would argue that systems analysis is value laden, depending on judgement, guesswork and intuition; that it is based on emotional interpretation of qualitative information; that it is concerned with constructive creativity in the grey areas of decision-making; that its approach is subjective and draws approval or disapproval from those involved; that it deals with areas of ambiguity which are open to misinterpretation; and, because people are involved, that it sets out to produce an acceptable conclusion rather than an accurate one.

The conclusion that one is forced to reach is that neither of these scenarios contain the whole truth. Systems analysis falls midway between an art and a science – in a similar way to management.

Exercises

2.1 Write down the type of communication which you would expect to pass between the systems analyst and all the people shown in Figure 7.

2.2 The manager of a local branch of a chain store which specializes in stationery, books, records, sweets etc. overheard a sales assistant explaining to a customer that it was the computer's fault that a particular book which he had ordered had not yet arrived. On making enquiries, the manager found that several of the sales assistants had this belief, even though it was quite erroneous. Explain why you think the sales assistants gave this explanation and what you think the manager should do about it.

2.3 You are a data processing manager who wishes to recruit a trainee systems analyst. You have decided to advertise the post to three groups of people:
(a) The existing programmers
(b) The staff in production control
(c) The graduates from a computer and business degree at a local polytechnic.
Draw up three fairly brief advertisements highlighting for each group the personality, knowledge and skills you are looking for, which they are likely to be able to offer.

2.4 Investigate the structure of a data processing department to which you have access. Produce an organization chart along the lines of Figure 8 and explain any differences between your chart and the one given in Figure 8.

3 Stages of system development

System life cycle

Systems have a common life cycle. They are designed, they are introduced, they evolve, they decay, and they are replaced. Whilst they are operating, they are susceptible to change from their environment and so continually need to be monitored and improved; eventually a state will be reached when it is no longer viable to patch a system or when it is no longer fulfilling its objectives, and then it will be necessary to carry out a thorough redesign; at this point minor modifications will decrease and a fundamental reappraisal will be carried out. Figure 10 illustrates this cycle. The upper sequence of boxes shows the operation of a series of systems ($X - 1$, X and $X + 1$). Above system X is shown its life cycle: there is a learning and adjustment period before it achieves optimum performance; it then operates efficiently for a period; and then because of internal or external factors it begins to decay and will after a time be replaced. In parallel to the operation of the system is the system development activity represented by the lower sequence of boxes; the bulk of this in terms of time is devoted to maintaining the evolving system as changes are required; the smaller part in time (but by far the larger in terms of resources) is concerned with developing a new system to replace the one which is in decline.

The life of data processing systems has started to shorten considerably. This is due partly to the rapidity of change in the business environment and partly to the continuing advances in computer technology. A data processing system which survives without major redesign for ten years is unusual.

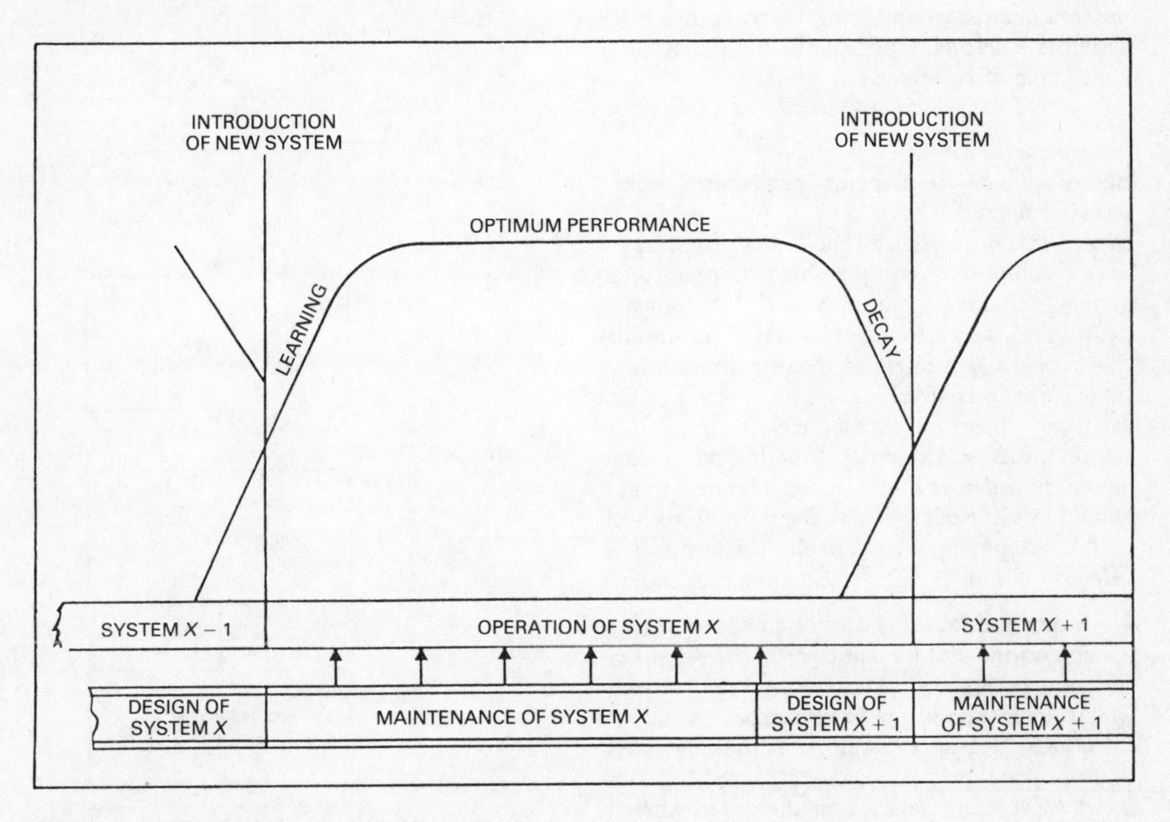

Figure 10 *System life cycle*

System development activities

The system development activity which was described above can be broadly divided into system design and system maintenance, but we need to divide the system design part into a number of stages concerned with feasibility, investigation, design and implementation. Thus the more detailed breakdown of the stages of system development activity might be represented as in Figure 11. Of course the activities of any particular project may not follow this sequence but in general the picture applies to most projects. It should also be noted that the stages are iterative to the extent that it may be necessary to go back to a previous stage for further work before continuing with the current stage.

Occasionally it may be necessary to purchase new equipment in order to operate the new system; in this case, alongside the detailed design and programming stage there will be two other activities – selection/ordering of equipment and installation of equipment.

Preliminary activity

Before a system development project can begin, terms of reference have to be drawn up. Ideally these terms of reference should stem from an overall computerization plan for the organization so that each development fits in with all other developments in a co-ordinated way. This implies that senior management of the organization is involved in decisions about computer development at an early stage and that an overall survey of requirements is carried out. Whether this is done across the organization or just in a limited area, it should be sufficient to determine the objectives of each development. For example, the terms of reference for any given project should include:

1 A description of the area of computer application and the anticipated benefits from computerization;
2 An indication of the likely impact on the organization of development in the area under consideration;
3 A suggested time-scale for the development with allocation of resources at an appropriate

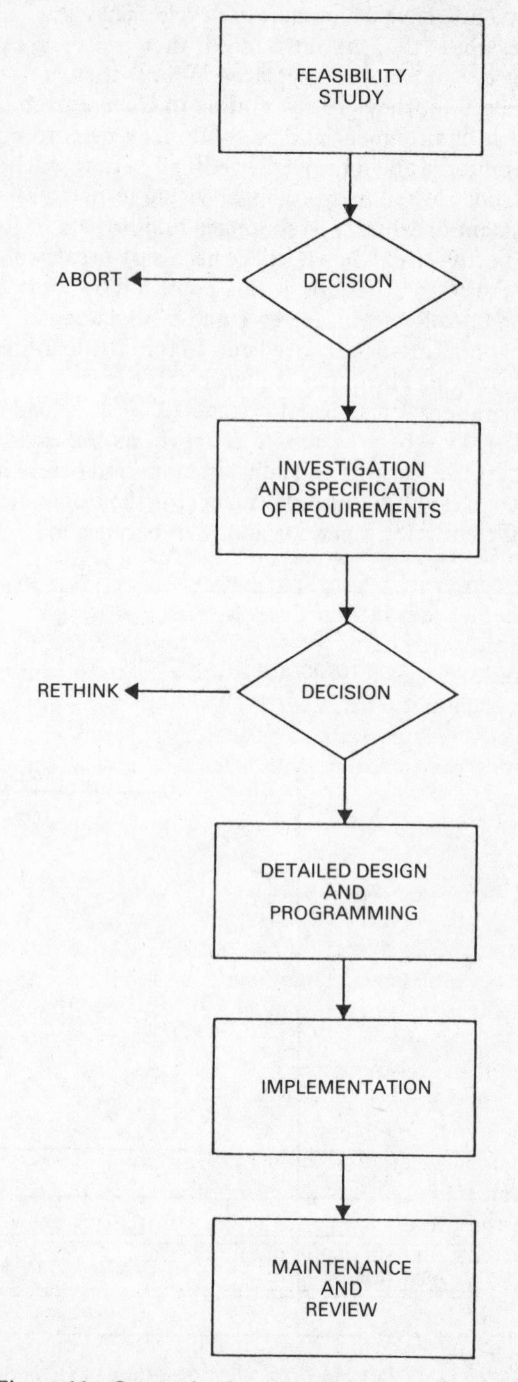

Figure 11 *System development stages*

level for the duration of the project; at this stage the time-scale/resources may only cover the feasibility study;

4 A detailed statement of the scope and limitations of the study to be carried out to try to avoid wasted effort on aspects of a system which are not to be affected.

This statement of terms of reference should come from senior management and be the guideline for the staff who will conduct the study.

Feasibility study

The next stage is to conduct a study of the feasibility of introducing a new system into the area under consideration. The feasibility study will be a complete systems analysis task in its own right; in other words it will involve investigation, analysis and design in outline, and it will produce a number of alternative solutions to the problem under consideration. Normally the feasibility study will be concerned with three aspects of the system – whether it is technically possible to achieve an improved system, whether such an improvement would be socially acceptable and whether it would be economically viable.

The technical aspect of feasibility will be concerned in outline with determining volumes of data and timings of activities and identification of requirements; these will be analysed and then formulated in terms of a number of possible approaches (with an indication of the hardware and software required) to meeting the requirements. Each approach can then be evaluated in social terms to determine its operational feasibility; this involves consideration of the effect on jobs, the possible organizational implications, and the effort required to prepare staff for the changes. Once a number of alternative solutions have been identified which are both technically feasible and socially acceptable, they can be financially evaluated. This involves identification of the costs of the present system, the costs of the proposed system, and the value of the benefits of the proposed system.

The feasibility study will result in a system proposal to management to allow a decision to be made about the preferred solution in the light of its financial return, its achievement of requirements, its social acceptability, its life cycle, and its relationship to overall development of the organization. Management must now decide whether to go ahead with developing a system, and, if so, with which system. If the alternatives are unacceptable, the project may be shelved or a more detailed investigation of feasibility may be instituted. The feasibility study will normally be carried out quite quickly – say, two weeks for a small system or three months for a very large system.

Investigation and specification of requirements

Once the decision to go ahead has been taken, a further set of terms of reference will be produced to authorize the systems analyst(s) to carry out a very detailed investigation of requirements. The flow of data, the forms and files used, the procedures, the equipment etc. of the current system will be documented in detail. Managers will be asked to define their objectives, areas of decision-making, and information requirements. These will be analysed and developed into a detailed specification of the user requirements for the new system (i.e. what output reports, file contents, input procedures, interrogation facilities etc. are needed).

This specification will not at this stage make any direct reference to the mechanisms (hardware or software) for achieving the requirements – rather it will be a detailed statement of what the users want the new system to do. This user system specification will then be subjected to a review by user managers before detailed design of the system takes place. At this review, the opportunity exists to stop the development if necessary – or, more likely, to refine the ideas until the specification is as good as the users can make it.

Detailed design and programming

Once the users have approved the specification of requirements it can then be turned into a detailed system specification. This involves decisions about which hardware and software to use, whether in batch processing or on-line processing mode, whether outputs should be printed or displayed, methods of data capture, and file organization and access methods. Output, files and inputs have to be specified; procedures (both clerical and computerized i.e. programs) must be defined; forms, codes and dialogues have to be designed; and provision must be made for adequate security and control in the new system. Clearly the detailed design stage is the most demanding in terms of staff and time.

The output from the detailed design will be a program suite specification from which the programmers will design and write the programs; a user manual which will tell the users how this new system is to be operated; and a computer operations manual which will specify the data preparation, data control and computer operating procedures. In parallel with detailed design, if extra equipment is required, effort will have to be devoted to selecting, ordering and installing the equipment.

The selection process involves:

1 Specification of equipment requirements – this will be passed to potential suppliers;
2 Receipt of proposals from the suppliers;
3 Evaluation of the proposals and selection of the one which most closely meets the requirements.

After the selection has been made, the order can be placed and then preparations made for the installation of the equipment. This involves a detailed examination of the equipment requirements (e.g. space, power supplies, ancillary storage etc.), site selection, site preparation (e.g. lighting, acoustics, safety, appearance etc.), and finally delivery and installation.

Implementation

Implementation is the most critical stage; it is the creation of a working system out of what so far has been a theoretical design. The emphasis now switches away from the systems analyst(s) to the users – and they will experience considerable unheaval, for which they must be prepared. Implementation planning needs to be carried out early and with full consultation with the users who will be affected.

The implementation itself involves training users in the procedures of the new system, converting files into a form in which they can be set up on the computer, testing the system as a whole, and changing over from the old to the new system. Most of these activities cannot commence until the detailed design and programming has been completed.

Maintenance and review

Once the new system has been implemented, it immediately becomes subject to change and so procedures are required to carry out amendments to the system to adapt it to changing needs. Most of the changes will arise as the system naturally evolves; some will arise from a regular review of the system carried out perhaps annually to assess whether the system objectives remain valid and whether the system continues to achieve those objectives in the most effective and efficient way.

Project documentation

As indicated earlier, the list of project stages is not definitive since each data processing department adopts its own approach to system development. Some will introduce more stages at a lower level of detail than has been described above; some will include programming under implementation; and some will omit the feasibility study in certain circumstances. But on the whole the system development process will be similar to that which has been described.

Out of each of the stages come various documents which have been briefly mentioned. It is perhaps appropriate now to bring these together in a diagram which shows the documents which a systems analyst might be involved in producing (Figure 12). In addition to ones already mentioned, it will be seen that during the three major stages (investigation, design and implementation) the team of systems analysts will also be required to produce progress reports to management. This is to facilitate control of projects which can often be of long duration.

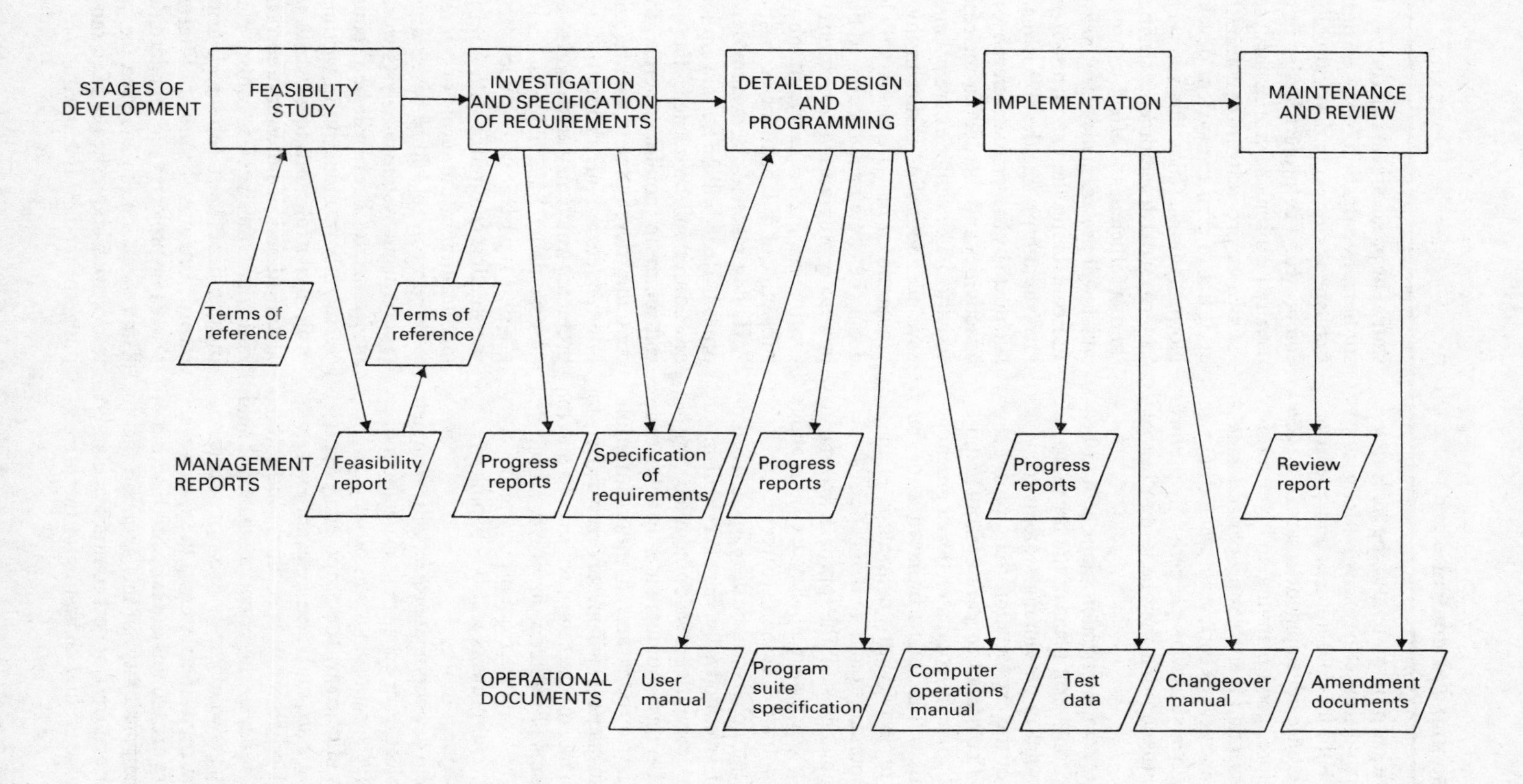

Figure 12 *Project documentation*

Project planning and control

Computerization projects tend to be high-risk ventures partly because they involve long time-scales, both for development and operation, and partly because development costs can be very high. Therefore, good planning and control of projects is essential; it enables proper assessment of priorities and ease of correction when things go wrong, and, in any case, people generally perform better in a situation that they believe to be well organized.

Long-term computerization plans should be based on the long-term plans of the organization; there is no point in developing systems for subsystems of the organization that are to be eliminated. The overall corporate plan should determine the overall computer development strategy. This will include a framework for the selection of projects (with a delineation of the subsystem boundaries and a clarification of interfaces), guidelines for the approach to be taken in developing systems, and policy decisions about equipment, software, system structures etc.

The long-term plans will be developed at senior management level normally via a mechanism known as the computer development steering committee. On this committee ideally will sit the heads of all the functional areas of the organization and the DP manager. It will meet fairly regularly to review the plans, to select the next projects and to lay down terms of reference for them, and to receive progress reports from project teams. Normally several projects will be ongoing simultaneously.

Each system development project will be carried out by a project team consisting of one or more systems analysts, representatives of user departments affected by the proposed system, programming staff, and other relevant experts (e.g. O & M officers, OR specialists, auditors, accountants). Ideally the project manager should be a line manager who can take executive decisions about the nature of the system. Often, however, line managers are too busy to take on this kind of day-to-day responsibility; in this situation the project may be run by a project committee, chaired by a line manager, and consisting of the project staff. The project team will carry out the feasibility studies, investigation, detailed design and implementation activities and will report to the committee. The project leader in this case will usually be a senior systems analyst.

Once the project team or committee has been set up it has to plan the system development. Normally project planning will set out to impose a pattern on the development activities which makes the most efficient and effective use of resources and which will provide a means of measuring progress and controlling the use of resources. Clearly this involves interpreting the objectives and constraints laid down by the terms of reference into a sequence of activities and then allocating resources to these activities. The activities are normally those which have been identified earlier in this chapter but broken down into a considerably lower level of detail (i.e. how many people to be interviewed, how many programs to be written etc.). Estimating the time required to carry out the activities is very subjective and dependent on past experience.

Having planned the system development project, work will begin and project control procedures will be needed. The aims of project control are to monitor progress, to keep the computer development steering committee informed, to provide an opportunity to sort out any problems which may arise, and to keep the project moving along. Sometimes as a result of project progress reports it will be necessary to go back to the steering committee with a proposal for modification of the terms of reference.

Project control will be based largely on progress reports from systems analysts working on the project team. These will be produced regularly (fortnightly or monthly) in writing as input to a project control meeting at which progress is discussed and, if necessary, corrective action initiated. Other project control mechanisms include time sheets (on which team members record their use of time on different aspects of a project or projects) and standard documentation. Time sheets can be analysed for quantitative information about the use of resources against the budget. Standard documentation, which all members of a project team are required to use to document investigations and design ideas, not only is useful to aid communication (as discussed in Chapter 9) but also can facilitate control by providing a visual check on what team members have been doing. Overall control (and planning) of project activities can be exercised by using techniques such as Gantt charts and network analysis, which it would be inappropriate to discuss in depth in a book of this nature.

Time-scale

It is perhaps appropriate to round off this chapter by placing the various stages of system development into context in terms of time-scale. It is impossible to generalize about time-scale because of the different sizes of systems and the different approaches used by different DP departments. The best one can do is to give an impression of the relative size and effort involved in the various stages. Figure 13 shows a profile of a typical project as far as implementation. The feasibility study is the smallest stage of the project in terms of both cost and time, though it tends to be prolonged by discussion of the alternatives and redrafting to meet management's requirements. The investigation stage is more costly and more time consuming because it needs detailed contact with user staff in order to iron out requirements. Detailed design and programming is the most costly and most time-consuming stage; large numbers of expensive computer staff are employed in producing the detailed specifications and writing the programs. Implementation is the shortest stage (invariably it has to be carried out over a very short period; otherwise two systems would need to be in operation simultaneously) but the most expensive in costs per day because large numbers of people are involved, especially in the user department. It can be seen that the two decision points for management (i.e. after feasibility study and after the specification of requirements) are quite crucial because the cost of each subsequent stage is much higher than the previous one.

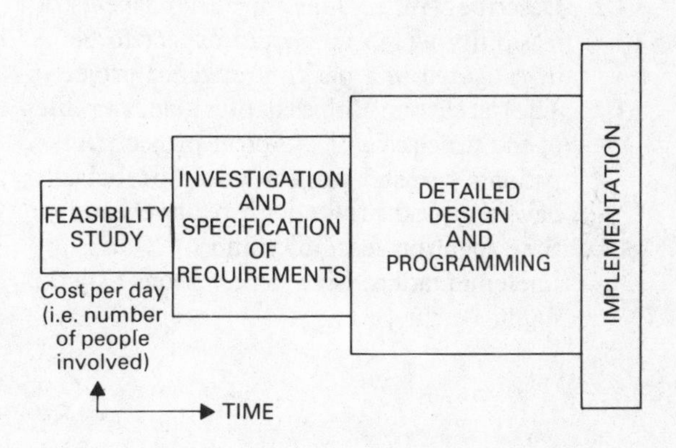

Figure 13 *Project time-scale*

It is, of course, not realistic to think of each of the stages as end-on partly because of the time in waiting for management decisions and partly because of changes in circumstances which can result in a switch of resources from one project to another. (Also, to a certain extent, the process is iterative; detailed design may involve further investigation, for example). Despite this, if one were to estimate the breakdown of time on a project whose total elapsed time was 52 weeks, one might expect about 8 to be allocated to feasibility, 14 to investigation and specification of requirements, 24 to detailed design and programming, and 6 to implementation. And if one were to break down the cost of developing a system, one might expect 5 per cent to be devoted to feasibility, 20 per cent to investigation, 60 per cent to detailed design and 15 per cent to implementation (bearing in mind that the cost of computer staff time is much higher than the cost of user staff time on average). This breakdown shows clearly the need for careful control of the investigation and design stages.

Exercises

An example of a data processing system in a hospital might be a patients' register on which is kept a record of all patients who have visited the hospital as either in- or out-patients, including details of their visits and a cross-reference to their detailed medical history.

3.1 Suggest some ways in which a computerized patients' register system might experience decay.

3.2 Describe with reasons the social aspects of feasibility which you could expect to be investigated in a patients' register project.

3.3 The last section of this chapter gives a profile of the time-scale of a typical project. A patients' register project would probably have a typical profile with relatively more time spent on feasibility and implementation. Give some reasons why this might be the case.

4 Feasibility study

Introduction

The first stage in developing a computer-based system is often a feasibility study; not all projects include this stage because some managers believe that their commitment to a development is so final that there is no point in using resources to test what is already a certainty. The objects of the feasibility study are to assess whether there are good technical, social and economic reasons for changing to a new system; to try to ensure that any new system which is developed will be acceptable to users, flexible in a situation of change and reasonably robust; and to produce a fairly accurate description of a proposed new system which can be used as a basis for negotiation within the organization and with manufacturers of equipment. Most experienced computer staff would emphasize that equal weight should be placed on technical, social and economic aspects of a proposed change, and that all the people affected by the change should have an opportunity to influence both the way the change is planned and the resultant system.

The feasibility study will normally be carried out by a project team reporting to the computer development steering committee. The team will consist usually of one or more systems analysts and a representative of the user departments affected by the proposed system (occasionally experts from other fields such as operational research or accountancy may be co-opted); but the work will involve staff from user departments as much as possible because they will have the knowledge of the existing system and views on what is required from any change. At this time some effort should be devoted to educating users about the likely impact of computerization (and the strengths and deficiencies of computers) to enable them to participate.

The basic activities involved in the feasibility study are investigation of current systems, outline design of possible new systems and evaluation of these alternatives. The work of investigation and design is described in the next two chapters and so will not be discussed here; this chapter will concentrate on the various issues involved in feasibility studies and approaches to evaluation.

Problems of feasibility studies

The most obvious problem of studying the feasibility of a system is that the study is concerned with the future, sometimes with forecasting ten years ahead. This is very difficult, especially in a rapidly changing environment, and yet any proposed system has to be evaluated in terms of its usefulness in the future rather than its immediate value. After all it will take quite a time to design and develop the system and, after implementation, there will be a period of learning when the system is not at its full potential. It might be as long as two years after the feasibility study before the new system is totally operational. A lot will have changed even in that period.

A second problem is deciding exactly how the feasibility is to be assessed. Normally senior management should identify certain objectives for the new system to fulfil; the difficulty here is whether the objectives can be measured, whether they are in conflict, whether they will change over a period of time etc. The criteria for measuring the achievement of objectives can be very subjective if they exist at all; for example, improved control, or better customer service or increased job satisfaction are notoriously difficult to measure.

A third problem is even more difficult for the feasibility team to resolve and that is the problem of which people should be considered to be affected by the system. To what extent, for example, should the general public be considered in a system for controlling town planning applications? Should the interests of trade unions weigh heavily in the evaluation of a personnel records system? Should customers have any influence on the design of a sales order processing system or suppliers on a purchase order system? Too often the feasibility study concentrates narrowly on the interests of managers without adequate consideration of the impact on other people.

Issues of feasibility

The feasibility team has to go through three stages in the process of feasibility assessment. The first might be described as 'problem definition'. This stage is largely concerned with investigating current procedures to determine the problems, requirements and opportunities. This involves the team in setting the boundaries of the system (always a difficult task); in trying to predict the changes that are likely over the life of the system; in examining the system as a whole in terms of its interfaces and information flows; in assessing the possibilities for improvements in the service to all the people affected by the system; and in identifying the alternative strategies which could be adopted to meet these opportunities.

The second stage is concerned with turning these alternative strategies into outline computer-based systems, which are more concrete and capable of being evaluated. An idea has to be formed of the approaches to gathering, storing and retrieving data, the equipment and people required, the volumes and frequency of activities, the timing of both development of the system and its operations; and the sources of supply of the various elements required.

The third stage is to evaluate these outline computer-based systems against the objectives set by management and to recommend one of the alternatives for detailed design and development. The evaluation will normally be concerned with three aspects – whether the outline system is technically sensible and viable, whether it is socially acceptable and whether it is economically beneficial. The approach is to rank the systems in terms of their technical and social compatibility, and then to evaluate financially the most acceptable sociotechnical systems.

Thus the major issues at each of these stages are technical, social and economic. These will now be examined in turn.

Technical aspects

The technical aspects of feasibility are largely concerned with answering the questions who? why? where? when? what? how much? and how often? in the context of both the existing system and any proposed systems. In other words an investigation of the present procedures is needed in order to identify the volumes, trends, frequencies and cycles of activity that will specifically affect the design of any computer-based system. It is this sort of quantitative data which enables more detailed costing of the proposed system. It also forms the basis for assessing the methods of input and data storage which are likely to be appropriate to the different types of systems.

Based on the findings of the investigation, various proposals can be put together for a new system; output reports, files, input methods and program requirements can be outlined, and their acceptability to the user gauged; need for specialist equipment or extra storage can be assessed; the implications for the operation of existing computer systems can be examined; the requirement for staff redeployment can be measured; and so on. For each of the alternative approaches to the problem, a system needs to be defined in outline which is technically feasible and which then can be evaluated both socially and economically.

An essential part of the technical feasibility study is the examination of different methods of developing the system and of running it. The system can be developed in a variety of ways: it can be designed and programmed within the organization; it can be contracted out to a consultancy firm; it can be developed in co-operation with other organizations which have a similar need; or it can be produced by modification of a bought-in software package. The cost of each of these approaches is of course quite different and they need to be carefully evaluated as part of the technical design considerations.

The data can also be processed in a variety of ways. It is not, for example, necessary to use one's own computer for processing; it might be more appropriate to use a computer bureau which sells computer time. Moreover, the processing can take place in either batch mode (i.e. transactions being grouped before submission for processing) or demand mode (i.e. each transaction being processed as it arises); the input of data can be on-line (i.e. under the control of the computer at the time of entry) or off-line (i.e. submitted to the computer having been first converted to computer acceptable media); the devices used can be either local or remote. All of these possibilities (together with the very wide range of alternative devices for input, storage and output of data) need to be taken into account in the outline design process. Often there is not a free choice about the type of system to be developed and operated because of existing constraints within the organization; however, if these constraints will lead to a less than adequate system, they should be explained in the feasibility report.

Social aspects

The social aspects of feasibility are largely concerned with the attitude of staff to the proposed change and the likely impact of the various alternatives on their jobs. This is very difficult to assess and is more the concern of user management than the systems analyst; it is the job of the systems analyst, however, to ensure that some attention is paid to the problem.

Some of the important things which need to be taken into account are as follows:

1 What is the level of knowledge of computerization among staff? Previous involvement in a systems project will help staff to understand what is happening and why. If this experience does not exist, then it is essential that some education is offered to staff as early as possible; this should cover computer appreciation in general and discussion of the proposed application area in detail.
2 How good are the mechanisms for consultation and discussion? Because a new system, if recommended by the feasibility study, is likely to lead to significant changes in the work situation of staff, it is essential that adequate channels of communication are set up if they do not exist. This is to facilitate airing of grievances but also to enable people to contribute ideas to the reorganization.
3 What is the organization's attitude to change? A major influence on individual attitudes will be his/her experience of previous changes; if consultation has been minimized in the past, then inevitably hostility will have been instilled in people's minds. This can only be eradicated by an open approach by senior management.
4 What will be the likely effect of the change on people's jobs? People tend to be happy with what they know and frightened of what is new. A careful analysis has to be carried out in the feasibility study of the impact of the various alternative solutions on individual jobs; need for redundancy, redeployment, retraining, removal etc. must be carefully anticipated even if only in broad terms. The introduction of a computer system should be seen as an opportunity to improve people's job satisfaction. This means attention has to be devoted to extrinsic factors like salary regradings, work environment, group relationships and social support; but equally important are the intrinsic elements of job

satisfaction such as autonomy, responsibility, challenge, variety. The social feasibility is concerned with examining how these requirements can be built into the new system.

Many of the points mentioned above are outside the control of the systems analyst conducting the feasibility study. Whether the needs of people are taken into account depends to a large extent on the authority structure of the departments concerned, and the leadership style of the most senior manager involved. If his style tends to be autocratic, then he is unlikely to be persuaded that staff should be either educated or consulted and he will probably dismiss the comments on job design as 'sociological claptrap'. There is little that the systems analyst can do other than to try to persuade the line manager of the validity of the approach.

Economic aspects

The feasibility study by this stage should have identified a number of technically and socially compatible systems and the requirement now is to evaluate these financially in order that they can be ranked for management.

In the economic evaluation, the team is concerned with comparing the costs of doing things in a particular way to the benefits. The evaluation will first gather information about the costs of the present method of operation; these break down into tangible costs (i.e. staff, accommodation, supplies, equipment, expenses (e.g. telephones) and capital tied up (e.g. in stocks, debts etc.)) and intangible costs (e.g. low staff morale, fraud, lost sales, poor cash flows etc.) and are calculated so that they can be compared with the cost of the proposed system. The costs of the proposed system on the whole are tangible and break down into the costs of development and conversion and the cost of operation. The costs of development and conversion include computer staff time, user involvement time, education and training, management and equipment and software acquisition. The costs of operation directly reflect the system which is proposed. Technically, the costs will relate to the scale of the system (volumes of data, frequency of processing, cycle of activity), the response time required, the location of operating units in relation to the computer processing facilities, the method of data capture selected, the security requirements, and the extent to which running costs of new equipment or software can be shared with other systems. Social costs will relate to the staff training and development needs, consultation mechanisms, salary changes, and the design of jobs.

Clearly costs will be affected by policy decisions of the organization to choose a particular manufacturer's equipment, or a particular piece of software; similar decisions about the processing

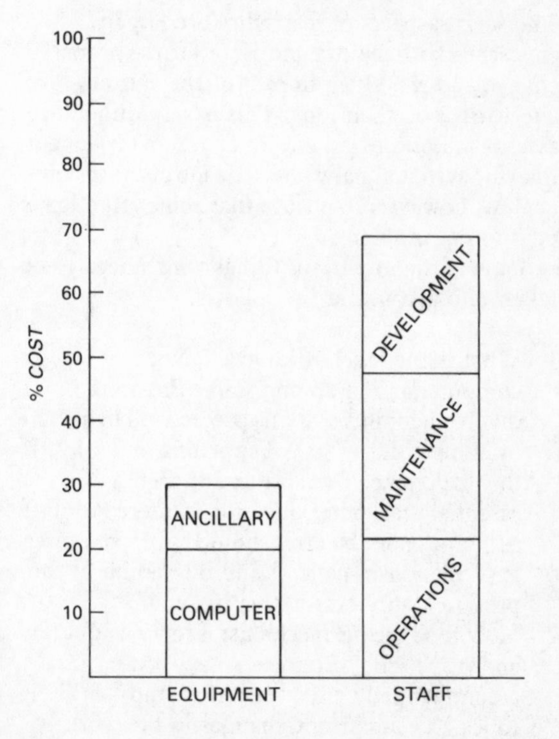

Figure 14 *Typical distribution of computer department running costs*

facility (bureau or in-house), the method of acquisition (purchase or rent or lease), and the approach to system development (in-house entirely or software house or package) will determine the costing.

An indication of the approximate relationship between the various costs of running a computer department is shown in Figure 14. This indicates that the major part of the cost is in staffing and not hardware; the cost of producing software is rising whilst the cost of hardware is relatively reducing.

The benefits of computerization tend to be broken down into two types – tangible and intangible. Tangible benefits are direct savings which can usually be easily evaluated. They include such things as reductions in staffing, accommodation and equipment, reduced stock investment, reduced maintenance costs, reduced peaking costs, avoidance of increases in numbers of staff, less staff turnover etc. These are clearly subject to a certain amount of guesswork but they are relatively easy to put a value on.

Intangible benefits (sometimes called gains) are more difficult to quantify. They include:

Better information for decision-making
Better planning (because of availability of corporate information)
Better company image
More control and discipline in systems
More flexibility
Better use of managers' skills
Quicker processing of data
Improved service to customers

These are benefits of computerization but they are difficult to evaluate; for example, the availability of better information doesn't in itself make a manager a better decision-maker, and improved service to customers is only useful if it is the right service (i.e. the right products are being offered at the right price). Generally, the feasibility team have to try to persuade managers to quantify the benefits which they feel they will gain from these improvements.

The benefits and costs of the new systems need to be presented in the form of a cost benefit analysis. The justification for expenditure on a new system must be that it will either produce more income or reduce expenditure. The alternative systems proposed therefore not only need to be practical and useful but also cost justified. The common way of presenting the cost benefit analysis is to show management a picture of cash flows involved in the system demonstrating the return on investment. The crudest method is the payback method where the benefits over a period are related to the cost outlay to show the profit over the period, the percentage rate of return, and the break-even point. This is shown in Figure 15. The clear disadvantage of this method is the fact that it takes no account of the declining value of money. A pound received now is worth more than a pound received in the future. More sophisticated methods such as net present value, discounted yield and discounted cash flow apply discounting factors to the income figures to gain a more accurate picture of the value of the benefits. Thus in Figure 16 the profit of £20,000 which was calculated in the pay-back method in Figure 15 is shown to be a loss of £5080 when a discounting factor of 10 per cent p.a. is applied. In other words, if money is losing value at the rate of 10 per cent per annum, a project costing £100,000 with an expected income of £120,000 over four years would make a net loss.

Cost outlay – £100,000			
Income	Year 1	£ 20,000	Profit = £20,000
	Year 2	£ 40,000	Rate of return = 20%
	Year 3	£ 40,000	
	Year 4	£ 20,000	Break-even point = year 3
	Total	£120,000	

Figure 15 *Payback method*

Cost outlay – £100,000			*Discounted at 10%*
Income	Year 1	£ 20,000	£18,180 (0.909)
	Year 2	£ 40,000	£33,040 (0.826)
	Year 3	£ 40,000	£30,040 (0.751)
	Year 4	£ 20,000	£13,660 (0.683)
	Total	£120,000	£94,920
	Profit	£ 20,000	–£5080

Figure 16 *Net present value method*

The net present value of the income is £94,920 rather than £120,000. This type of approach allows the organization to compare all its capital projects on a similar basis. (Various other techniques can be used to measure risk, handle probabilities, introduce forecasts of inflation, and take into account investment grants and tax allowances etc., and the process can be very sophisticated).

It is therefore worth pointing out that, although the systems analyst is unlikely to be involved in detailed calculations of discounted cash flows, since the accounting department will tend to handle all projects on a common basis, the accuracy of the figures which he supplies to the calculations is of paramount importance. The validity of the cost benefit evaluation depends not on the techniques of manipulation of figures but on the accuracy of the original estimates.

Feasibility report

The feasibility report must reach some conclusions about the proposed system even if the conclusions are that there are two or three possible approaches to be followed. It should cover these areas:

1 Description of the area of activity under consideration, the objectives to be satisfied and the relationship of the development to the overall plans for computerization.
2 Description and specification of the existing system, its problems and advantages, and the requirements of the new system. This section should also give figures on the costs of the existing system for comparison purposes.
3 Description of alternative proposed systems (including technical and social aspects) in terms of how they will work, how the organization will be affected, and how much they will cost.
4 Evaluation of alternatives and recommendations.
5 Development costs and time-scale of development (including a discussion of critical factors). (This section of course is not necessary if the recommendation is for no system to be developed.)

The evaluation in area 4 is the key part of the report and should cover all aspects of the system alternatives and not just costs and benefits. Some discussion should be presented on the security aspects of the system (e.g. reliability, accuracy, quality control), on the usefulness period (e.g. capacity for expansion, flexibility, maintainability), on the user reaction and degree of user management support, and on the time-scale of development, implementation and operational life. In particular the benefits should be carefully analysed, with some specific estimates of the likelihood of their being achieved.

The report will be submitted to the computer development steering committee and a decision

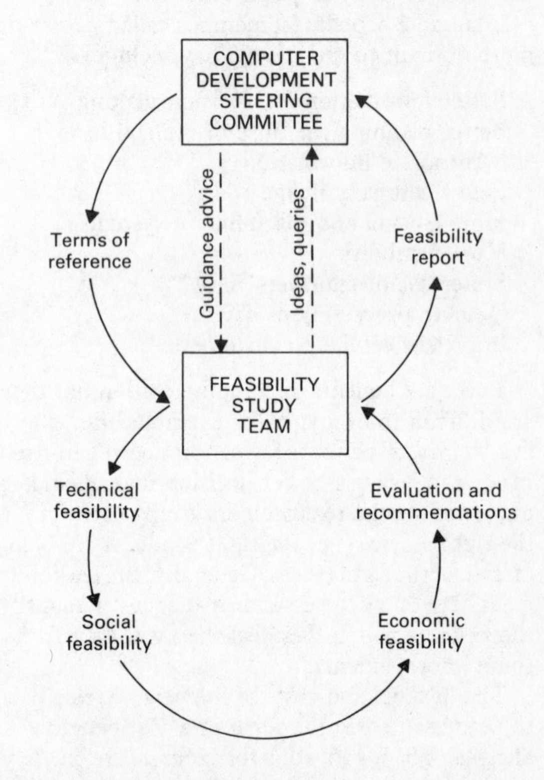

Figure 17 *Iterative nature of the feasibility study*

will be made on the approach to be followed. This is the major decision point in any project because of the high expenditure on system development which follows the go-ahead. The whole process of feasibility is iterative, however, and it might well be that the steering committee will reject the recommendation or ask for further investigation to be made. Figure 17 illustrates this iterative process. One final point, which perhaps should have been made earlier, is that the proposed approaches do not necessarily have to involve the use of a computer. It is perfectly reasonable for the team to recommend an improvement in the system without introducing computer methods.

Exercises

4.1 There is a proposal to introduce a computerized pupil record for all schools in a local education authority. Identify the people who you consider are likely to be affected by the system, and for each explain with reasons whether you think they should participate in a feasibility study and what their contribution might be.

4.2 A variety of different approaches to developing and running a computer-based system are described in the section 'Technical aspects'. Describe for each one conditions which might be favourable for the adoption of the approach.

4.3 In Exercise 1.3 you described the benefits of three computer applications. Comment on the extent to which the benefits which you identified can be quantified.

4.4 A new system will cost £150,000 to develop. The net income over the estimated five years of its operational life is:

Year 1 £13,000
Year 2 £35,000
Year 3 £35,000
Year 4 £70,000
Year 5 £75,000

Calculate the net present value (NPV) of the development using discount rates of 10 per cent, 12 per cent and 14 per cent and make a recommendation to management.

The discount factors are:

Year	at 10 per cent	at 12 per cent	at 14 per cent
1	0.909	0.893	0.877
2	0.826	0.797	0.769
3	0.751	0.712	0.675
4	0.683	0.636	0.592
5	0.621	0.567	0.519

It would be possible to reduce the development cost by £10,000, but this would mean no income in year 1. What effect would this have on your recommendation?

5 Investigation

Introduction

The investigation of an existing manual or computer-based system will begin when the computer development steering committee has decided what the investigation is aimed to achieve. It may be that the investigation initially will be part of a feasibility study intended to provide information to allow some assessment of the viability of a project – or it may be that the investigation is to be conducted in depth as a result of the recommendations of the feasibility study. In either case, the investigation team should be provided with terms of reference. Ideally these should be negotiable so that they can be agreed by all the participants. The terms of reference will define the scope, the objectives, the constraints and the resources of the investigation.

The *scope* will often be defined in departmental or procedural terms (e.g. investigate the sales ledger activities of the accounts department); this can be awkward when the investigation of information flows inevitably needs to cross organizational boundaries, but the intention is to provide boundaries for the investigation. The *objectives* should be expressed (ideally) in a quantifiable way, but often they are rather vague; it is very difficult to measure the achievement of an objective such as 'to provide management information', whereas an objective such as 'to reduce the time delay in sending out invoices from one week to one day' is more concrete and measurable. It is not unusual, however, for objectives to emerge during the investigation as more knowledge is gained about the problems of the department under review. The *constraints* should pin-point those aspects which are not going to change; for example, if no extra equipment is to be purchased, this should be known from the start, or, if no staff are to be made redundant, this should be publicly announced; there is no point in the investigation team wasting time on possibilities that do not really exist. Finally, the terms of reference should include a statement of *resources* available for the investigation, i.e. how many staff for how long.

The decision to investigate a particular aspect of an organization's activity will often emerge from rather vague feelings among managers that there is a problem. It may be felt that service to customers is deteriorating or that too much capital is employed or that cash flows need speeding up to improve liquidity. In each case a variety of approaches can be taken to the problem and the systems analyst may find himself having to identify the most appropriate. This clearly involves a reasonably wide study of the organization as a whole and a mature experience of the nature of business policy and organization. In this type of exercise, terms of reference will be less specific.

Approach to the investigation

Once the project team has been given the terms of reference, the investigation has to be planned in some detail. The first requirement is to build up background information about the area of study to enable identification of specific problems, procedures and people which need thorough coverage. Once the team knows what the investigation is going to involve, it can then plan interviews with the appropriate staff to discover details of the type of information required, the current procedures and their deficiencies, and the files, documents and records used.

The approach to the investigation must always be formal – working down the hierarchy from the most senior manager. It would be discourteous to conduct interviews with junior clerks without having first sought the approval of their supervisor and his manager. The systems analyst should also deal with staff in a confidential manner so that staff feel they can trust him and talk openly about the problems as they see them without being 'reported' to their superior. Part of this atmosphere of trust needs to be created by honest and free discussion of the implications of the investigation, and by keeping people informed of the way things are developing. And, of course, the help that users provide to the systems analyst must be gratefully acknowledged.

Throughout the investigation it is valuable to encourage users to participate in the assessment of requirements and problems of the system. The more they can be involved in and contribute to the systems analysis activity, the less will be their alienation from the eventual findings. In any case the users have far greater knowledge of the operation and shortcomings of the existing system than the systems analyst.

Background information

It is necessary for the systems analyst to gather some background information before the investigation proper can begin. This information is required to enable the planning of the detailed investigation and also to provide talking points when interviews with user staff are under way. A lot of background information can only come from work experience in the organization and talking to other members of staff. Thus a systems analyst who has worked for an organization for a long time will need to spend much less time picking up background information than the new recruit.

The kind of background information that is needed breaks down into four main areas:

1. First, there is information about the environment in which the organization finds itself; here one might wish to know about the competitive situation, relationships with other similar organizations, position *vis-à-vis* customers and suppliers, impact of government policies, labour relations climate etc.
2. Second, there is information about the organization itself, its policies and the atmosphere within various parts of the organization; this information can be gathered to some extent from company reports and trade journals but mainly from informal conversations with other employees.
3. Third, there is information about the structure of the organization, e.g. the number of departments/divisions, their location and staffing levels, and the nature of their responsibilities. Equally important is some feel for the extent of planning within the organization and the style of management adopted. Detailed information is needed about the formal organization structure of the area of the organization which is being investigated, its operational activities and terminology used.
4. Fourth, valuable background information is provided by previous investigations in the area under consideration, conducted either within the organization or in other organizations; these might include work study, O & M and job evaluation exercises, or descriptions of packages used for particular computer applications.

This background information needs, of course, to be specific to the particular area under investigation. For example, if one is concerned with a production control system, one might be interested in knowing about raw material supplier performance, labour relations, absenteeism, payment systems, government energy policies, trends in product demand, the structure of the production department, number and responsibilities of staff and their relationships with other departments, attitudes of managers and staff to computerization etc. This sort of information would determine the initial emphasis of the investigation. It would help the analyst to form an idea of whom he would wish to interview, in what sequence, and what he would wish to seek information about. The background information will enable him to talk reasonably intelligently to specialists and to formulate some ideas to organize a discussion with senior managers. The relative importance of different pieces of information can only be gauged as experience is gained.

Detailed information

When the systems analyst begins the investigation in detail he will tend to find himself gathering information at two levels – management and operational. At management level, the requirement is to discover what decisions the manager makes and what information is needed to enable the decision to be formulated and made; at operational level, the investigation will be concerned with the various procedures that are carried out, the data (in files, on forms etc.) that is used, and the information which is created for use by management.

Management level

There are various levels of management with organizations; they are usually categorized as top or strategic, middle or tactical, and junior or supervisory. Depending on the scale of the investigation, all levels or perhaps just one level will be involved. Regardless of the level, the systems analyst needs to discover how a manager approaches his task – how he plans his work and the work of his department, how he organizes and staffs the tasks that have to be performed, and how he measures performance. He needs also to identify the decisions that the manager is required to make and the information he uses (or requires) to support the decision-making. Most importantly the analyst has to try to determine what the manager's objectives are. Often these are difficult to elucidate when managers are not used to thinking in terms of objectives, but identification of objectives is very helpful in determining information needs. Information is required to enable the manager to plan to achieve his objectives and to measure the achievement.

In addition to the manager's own requirements, it is necessary to discuss with him the way in which his department operates and the good and bad features of its operation; in particular, one might wish to discover the bottlenecks and constraints that system redesign might overcome.

Operational level

At operational level, the systems analyst must attempt to find out the detailed procedures of a department i.e. who does what, when, where, how and why? This information needs to be comprehensive and thorough, covering all the different types of activity. Accommodation, supplies, equipment all need to be examined to assess their contribution to the procedures. In particular, the systems analyst needs to discover the difference between routine, control and exception procedures. Often the exceptional procedures (such as handling errors, priority orders, and dealing with peak workloads) are forgotten by the clerical staff when describing their system; and usually the exceptional procedures, which are relatively simple to build into clerical systems, are expensive to build into computer-based systems. Problems which emerge from the investigation (such as high error rates, excessive duplication of activity, faulty control procedures) should be thoroughly examined to ensure that they are understood by the analyst.

Each procedure will make use of lots of data and the analyst has to identify the data involved. It may be in the form of documents, verbal messages, or information kept in the user's head. Files of documents, which have accumulated over a lengthy period, are useful to the analyst; and master files (e.g. customer records, supplier records) are extremely important as the reference points around which procedures are built. In particular the analyst will want to discover the volumes of documents over a particular period (and the rate of growth), the frequency of usage, and the size of master files. This quantitive information is essential to the job of sizing a new system.

Information flow

The previous few pages have tended to suggest that the investigation should be confined to one department, or at least within the boundaries of a number of departments, but in reality, of course, information flows across departmental boundaries. It is frequently appropriate therefore, to conduct an investigation of information flow rather than departmental procedures. This can be done in a variety of ways:

1 One can take a particular input document (e.g. an order) and chart its progress through the organization, recording how it is used and what happens to the data which is on it;
2 Another approach is to take an output document (e.g. a VAT report) and analyse the procedures and data that are required for it to be produced;
3 A third method is to follow physical flows (e.g. of material, or people, or products) and to analyse the information requirements at each point of the flow.

In all three cases the object is to achieve a total view across the organization rather than a partial, sectional view of one department.

Investigation methods

Having identified the kind of information that needs to be gathered in the investigation, the analyst has to choose the best methods of finding it. The methods which are commonly used are interviews, questionnaires, observation and searching through records. Information from managers can only be gathered by interview; information about procedures can be gathered by a variety of means.

Observation

In the course of an investigation, visits have to be made to the departments to talk to staff, look at documents etc., and these visits can be used to pick up information about the environment of the system by observation. Often, problems in existing procedures stem from poor environmental conditions. For example, there may be too much noise; the lighting may be poor; the temperature and ventilation may be conducive to muzziness; the furniture or equipment may be inadequate; the layout of the offices may cause too much movement or prevent easy access to files; the conditions of work may be quite unpleasant. In addition to observing working conditions, the systems analyst may be able to assess the smoothness of the flow of work, the evenness of workload, the closeness of supervision, and the number of interruptions. A lot can be learned also about the pace of work and methods used, just by watching. This is not to suggest that the systems analyst should spy or even sit inspecting the department's work; rather the requirement is for him to be observant when visiting the work stations to conduct interviews or to examine records. Any points which are noticed in this way should obviously be followed up in detail in the interviews.

Record searching

The purpose of searching through existing records is to establish quantitative information about the data and procedures in an existing system. For example, a count of the customer index cards will indicate how many customers the organization has; examination of the dates on which they first placed business will provide information about the average life of a customer, the number of new customers per month etc. An inspection of amount owed on individual customer account records will indicate to the systems analyst the number of characters required to store 'amount owed' per customer on file; and so on. In addition to the establishment of quantitative data (volumes, frequencies, trends, ratios etc.), the searching of existing files helps in checking points that have come out of interviews and identifying points that have been missed in interviews. For example, exception situations which tend to be forgotten by a user in the course of an interview usually come to light in a detailed search of document files etc. A particularly useful source of information is found in existing descriptions of procedures. Sometimes these exist in notebooks kept in drawers etc.; less frequently they are formally typed and kept by the supervisor or manager; occasionally they are kept as part of a job description that has been written for the personnel department when a new employee has been recruited.

There is a considerable amount of quantitative information which cannot easily be gathered by searching existing records. Examples are the number of telephone calls to a particular office at various times in the day (e.g. for phoned orders) or the number of times a file is accessed during a day (e.g. to check the credit-worthiness of a customer who has placed an order) or the number of enquiries from various parts of the organization about the status of a particular file (e.g. the quantity in stock of a particular product). (All of these are examples of essential quantitative information which will influence the design of a system.) In these situations, the normal practice is to use special records to gather the required information and the most commonly used method is a tally sheet (see Figure 18). On this the user places a mark in the appropriate column to record the occurrence of an event. The important thing about tally sheets (or any special purpose record) is that they should be simple and used for a limited duration (because they cause additional work for the user).

Number of phoned orders received

Day \ Time	0830–1000	1000–1200	1200–1400	1400–1730
Monday	卌 卌 卌 卌 III	卌 卌 II		
Tuesday				
Wednesday				
Thursday				
Friday				

Figure 18 *Example of a tally sheet*

Questionnaires

Questionnaires are the most difficult method of fact-finding to use successfully. An enormous amount of time has to be spent on questionnaire design in order to ensure that the replies are valid. The use of a questionnaire should therefore only be attempted after careful consideration, and when the systems analyst is faced with gathering information from a large number of people or from a number of remote locations. For example, an investigation of library cataloguing might justify sending a questionnaire to the many thousands of library users; or a study of sales order processing procedures at 100 different branch offices might require a questionnaire to establish the degree of similarity or difference in the procedures. But the replies to a questionnaire should always be regarded with suspicion because it is so difficult to use words within it which will be objectively understood and interpreted in exactly the same way by all of the respondents.

If a questionnaire is used, it should be very carefully designed. The *purpose* of the questionnaire should be explicitly defined and explained to the recipients in a covering letter. The *recipients* should be carefully chosen; if the right person to complete the questionnaire is an accounts clerk, then it would be inappropriate to send it to the accounts manager; rather, the latter should be asked for permission and receive an explanation of why the questionnaire is going to his subordinate. The *questions* should be phrased very precisely in a way that the recipients are in no doubt about their meaning; ambiguity is one of the most common features of questionnaires. There should always be space for the recipient to answer each question; even in a multichoice situation where you feel you have covered all possibilities, there should be a space for 'other (please specify)' to cover exceptions. The *period* to which the questionnaire relates should also be carefully defined; often people's replies will differ depending on the time-frame of their answer. Equally important is the *timing* of the questionnaire; it should arrive at a time which is convenient to the recipient, it should relate to a specific point in time, and replies should be requested within a certain period. Some *incentive* to complete the questionnaire should be offered; sometimes a free pen may be enclosed, but frequently the best incentive is an enclosed, stamped and addressed envelope.

Follow-up is very important for those questionnaires which have not been returned; the follow-up letter should be prepared at the same time as the questionnaire and be ready for despatch perhaps a week after the replies are due. Finally, all questionnaires should be given a *trial run* with a sample of respondents to check that the questions are understood and are not ambiguous.

Interviews

Interviews are by far the most important and most rewarding method of investigation and for many aspects (such as requirements, problems, inadequacies of the existing system etc.) are the only method of finding out what is wanted. They must be carefully prepared, smoothly conducted and adequately recorded.

Planning the interview

The first question that needs to be answered in preparing for interviews is *who* is to be interviewed. This is simple to answer because you must interview those who will provide the relevant information at the appropriate level. Two things should be borne in mind; one is that the interviews must follow the structure of the organization, starting with the most senior person and working down the hierarchy (out of politeness); the other is that, although a detailed plan of interviews is needed in advance, it should be treated flexibly to allow for changes as new information comes to light and the need to see other people is recognized. The *sequence* of interviews obviously needs to be the one which will allow the best build-up of information, but often it cannot be followed because the interviews have to be conducted at times convenient to the interviewees. The *location* of an interview should be chosen to allow quietness and concentration, but it is probably best to conduct it in the interviewees's own office, partly because this will encourage him to feel more at home, and partly because it provides an opportunity to observe what is going on in the office; also interruptions during an interview can reveal aspects of the system which might otherwise be missed. Having planned the interview sequence and location, the analyst finally must prepare for the interview by deciding what he wants to find out and constructing a *checklist of questions* to ask; this checklist should be a guide for the interviewer rather than a fixed list which will be followed at all costs. The questions should be at a level appropriate to the interviewee; managers should be asked about policies, decision-making, responsibilities and requirements; clerks should be quizzed about forms and procedures at a detailed level.

Conducting the interview

The systems analyst should arrive for the interview promptly at the prearranged time and should take a formal approach to the interview in terms of address and questioning. The interviewer needs to introduce himself, explain why he has come and what the purpose of the interview is. Permission should be sought to take notes during the interview and copies of any documents or descriptions of procedures should be obtained. The interview needs to flow quite smoothly and so it should commence on ground that is familiar to the interviewee. The questions should be relevant and at the right level, and each point which arises should be followed up in detail until the analyst feels that he understands the facts and the interviewee's views of the topic. The essence of a good interview is that the interviewee should feel at ease and feel that the interviewer is interested in his problems. The systems analyst needs to sit back and listen, allowing the user staff to talk freely about their jobs, but at the same time ensuring that the interview covers the ground which is needed. Argument and criticism are quite inappropriate to the interview situation; it is the analyst's task to listen to everything that emerges and to judge both its relevance and its validity; at times other (possibly conflicting) viewpoints will need to be described, but this is to enable more light to be thrown on the situation rather than to test the particular interviewee's competence or sincerity. What is needed is an atmosphere of trust.

During the course of the interview, the systems analyst must ask encouraging questions and keep the discussion flowing by nods of understanding and agreement. Terminology must be correctly interpreted on both sides and any doubts cleared up at the time they arise. The analyst should steer the interview in the direction he wants it to go but not apply too much pressure in areas about which the user is clearly unwilling to elaborate. Simple questions, asked one at a time, are the most useful; questions which are biased or rhetorical or leading should be avoided. Interviews are not simply intellectual, verbal exercises; much is to be learnt about the interviewee's views from intonation, facial expressions, eye movements, gestures etc.

At the end of the interview the systems analyst should always thank the interviewee and try to leave the door open for a return visit if necessary.

After the interview

As soon as possible after the interview, the notes which have been taken should be knocked into shape in a formal report. Occasionally the interviewee will ask to see the interview report, and, in an effort to check the accuracy and completeness of the interview record, it is useful for the interviewee to be asked to read the document. Also important is the need to cross-check the interview with the findings of previous interviews to ensure consistency. At the completion of each interview, the overall plan should be reviewed to ensure that it is still on target. At any interview, information may come to light which causes a change in tactics.

Recording the investigation

It is essential that the findings of the investigation are adequately recorded for four reasons:

1 The systems analyst cannot retain all the information in his mind, and so he needs to commit things to paper;
2 Often several people will be working on the system investigation simultaneously and all will need access to the information;
3 Often it is necessary to confirm or check findings with users and they will need a written record;
4 The findings need to be analysed and mulled over for a period and the act of recording helps the process of analysis.

In order to assist with these various requirements, it is desirable to use a standard form of documentation on all computerization projects with very specific guidelines on what is to be recorded and how. The guidelines need to cover four types of information – unstructured findings, procedures, data and relationships.

Unstructured findings

The most obvious type of unstructured findings are the opinions expressed by users during the course of interviews (e.g. management requirements for the new system, problems with the old etc.). These need to be recorded as

Discussion Record NCC	Title	System	Document	Name	Sheet
	Interview with Sales Office Manager	SOP	2.1	INT-SOM	1/1

Participants
F.R. Brown, Sales Office Manager
J.G. Lewis, Systems Analyst

Date 19.1.83

Location
Room S4.7

Objective
To clarify the procedure for updating the outstanding orders file

Duration
20 mins

Results:	Cross-reference
Mr Brown confirmed that:	
(a) the outstanding orders file is reviewed twice weekly – on Tuesday and Friday afternoons.	
(b) on each occasion the file is searched for any item which is outstanding which appears on the latest <u>Goods Received List</u> from Purchasing.	SOP/4.1/GRL
(c) if an item is found, the order is removed from the outstanding orders file and <u>processed</u> as though it is a new order	SOP/3.2/ORD

Figure 19 *Example of a completed S21 Discussion Record, recording the results of an interview*

narrative on a special sheet which enables the content to be defined and cross-related to other documents such as flow charts and data descriptions. The sheet should have space for the date of the meeting, the names of the participants, and the location – and all facts recorded should be cross-related if possible in a margin. An example of such a form is the NCC Discussion Record (S21) – see Figure 19.

Procedures

Procedures are usually defined using structured narrative, charts/diagrams or decision tables. The use of each of these is described at greater length in Chapter 10. Just a few observations need to be made here.

Structured narrative can be used to define any procedure.

Decision tables will tend to be confined to those parts of a procedure where decisions are made based on clearly defined rules and where the variables affecting the decision are relatively few. For example, the decision in sales order processing as to whether an order is credit-worthy may be one which can be well defined by a decision table.

Charts and diagrams are drawn at various levels

Clerical Document Specification NCC	Document description: Goods Received List	System: SOP	Document: 4.1	Name: GRL	Sheet: 1/1
	Stationery ref.: None	Size: A4	Number of parts: 2	Method of preparation: Typed /using carbon	
	Filing sequence: Date	Medium: Ring binder		Prepared by: Goods Inwards Clerk	
	Frequency of preparation: Daily	Retention period: 12 months		Location: Goods Inwards Office	

VOLUME	Minimum	Maximum	Av/Abs	Growth rate/fluctuations
	1 sheet per day	10 sheets per day	4 sheets per day	Approx 3% per annum

Users/recipients	Purpose	Frequency of
① Top copy to Sales Office	For checking against outstanding orders file	each Tues and Friday
② Carbon Copy to Goods Inward File	For checking (if needed) at some later stage	—

Ref.	Item	Picture	Occurrence	Value range	Source of data
1	Title	x (19)	1	GOODS RECEIVED LIST	Goods Inward Clerk
2	Date	x (8)	1	99/99/99	
3	Product Code	9(6)A	40 per sheet	00000/A-69999Z	
4	Quantity Received	9 (4)	1 per product code	0001-9999	
5	Signature	—	1	—	

Notes

S41

Figure 20 *Example of a completed S41 Clerical Document Specification, recording details of a clerical document, the goods received list*

during the investigation stage. An overall data flow diagram (described at the end of this chapter) provides an overview of a system. As more detail of a procedure becomes known a clerical procedure flow chart defines the detailed activities at each point in the system. As the charts are constructed and each symbol is tested for its meaning, relevance and need, the analyst gains an analytical understanding of the system.

Data

In investigating existing procedures, the systems analyst invariably collects copies of all of the forms etc. used in the system. These are essential to his understanding of what is going on; ideally he should ask to see them completed. A lot of the information on the forms is self-evident but extra information may need to be gathered about the data. This kind of information might be:

Method of filing
Retention period for the document
Volume of documents used per period (including estimates of growth)
Distribution of the document and its use
The size and format of all the items of data on the form, the possible code values which can be entered, and the source of the data

This sort of information can be collected during interviews and should be recorded on a form such as the NCC Clerical Document Specification (S41) – see Figure 20. The quantitative information about data should also be recorded in the form of tables and graphs – this will be useful later in sizing the new system.

Relationships

Relationships between events, data and people often need to be documented and other methods are available for this. For example, relationships between items of data are shown on grid charts (see Figure 21); relationships between people are often shown in the form of an organization chart (see Figure 22); and physical movement charts can be used to show how documents pass around an office or people move about in an office (these generally take the form of an office layout drawn on grid paper with the movements of forms/people shown as lines on the layout).

An organization chart shows the relationship between people at different levels in the hierarchy. These are formal charts and do not necessarily reflect the real positions of power or the real lines of communication. Though they can be quite misleading, they do provide a shorthand way of recording the organization. They tend to be rather political documents and so they should always be dated and include a rider that position on the chart doesn't reflect relative seniority.

Grid charts are very simple charts used to compare and relate different factors in an investigation; for example they can be used to analyse which items of data are common to several forms, or which decisions are made by which managers, or which decisions use which information. They tend to be more of an analytical tool than a recording tool.

DOCUMENTS / ITEMS OF DATA	Customer order	Picking list	Invoice	Stock record	Customer record		
Customer no.	√	√	√		√		
Customer name	√	√	√		√		
Delivery code	√	√	√		√		
Product no.		√	√	√			
Product name	√	√	√	√			
Quantity ordered	√	√	√				
Order date	√	√	√				
Customer order no.	√	√	√				
Our order no.		√	√				
etc.							

Figure 21 *Example of a grid chart which relates documents to the items of data which appear on them*

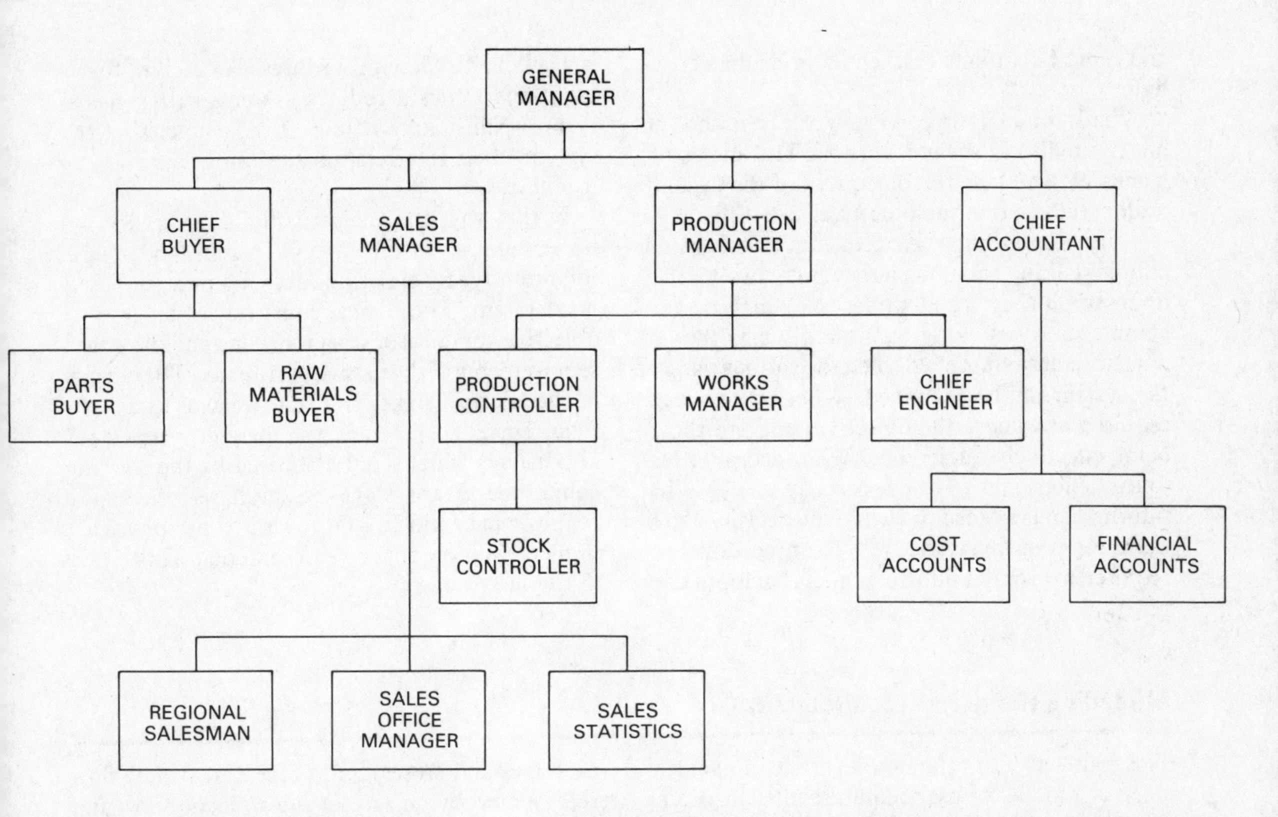

Notes: 1 This chart was correct at 24 June 1980. 2 There is no significance in the relative levels of the boxes in the chart.

Figure 22 *Typical organization chart for a manufacturing company*

Analysing the findings

The analysis of findings is not a separate stage of the overall activities but an integral part of the investigation process. The analyst is continually thinking over the implications of what he has discovered and trying to assess the possibilities of changing the existing system. All the time he is trying to measure his findings against the original objectives set for the investigation (and also against the requirements of the users). He doesn't have a completely free hand in designing a new system; there are considerable constraints of time and money and user acceptance. In particular the analysis has to take place within the given organizational framework of policies, objectives, structure etc. But despite this, the analyst has an obligation to question everything which he is told and to test its validity.

The approach to analysis is normally to examine what happens in a system (and when, where, who and how) and to ask why this happens in this way and whether it could be improved by happening a different way. The analyst will discover symptoms of problems (such as poor communication between staff, high absenteeism rates, excessive copying from one document to another) which will point him in the right direction in looking for improvements, but essentially he must take a wider approach, often seeking more fundamental and radical solutions than come immediately to light.

A basis for analysing a system can be found (not unexpectedly) in systems theory. The analyst begins by assessing the objectives of the system under study; often the objectives are difficult to identify because the managers concerned are happier talking about what they do than why they do it; when they are identified they may be very complex and not easily quantified. Once the objectives are established, then the various parts of the system can be examined. What actions are required to achieve the objectives and are they being effectively carried out? What decisions lead to the actions and are they correctly made? What information is needed to enable the decisions to be made (or even formulated)? What procedures are required to produce information? What input data is available to these procedures? As each of these questions is considered, the systems analyst has to look in painstaking details at the elements of the system, their interrelationships and their usefulness.

In this way, the analyst is building up an understanding of the current system, and, more importantly, formulating ideas about a proposed new system. These ideas must reflect his own objective appraisal of what is needed and the stated requirements of the user departments. The output of the analysis stages is a statement of user requirements, usually in the form of a series of alternatives which can be debated by the steering committee. At this stage the emphasis is on logical requirements and has no regard to the physical implementation of the requirements. This comes in the next stage.

Modelling the users' requirements

The main output of the investigation and analysis stage is a model of user requirements. In recent years, with the advent of structured approaches to systems analysis and design, the main tool for modelling has been the data flow or activity diagram. (The conventions for drawing a data flow diagram are described in Chapter 10.) A data flow diagram is intended to provide a unified approach to analysis and design by concentrating in a structured way on the logical requirements of the system. Its perspective is the flow of data between the processes which transform the data, rather than the location, control or physical implementation aspects. It is based on a hierarchical, top-down decomposition of the logic or functions of the system. The structured approach can be taken through to the design stage because the lowest level processes can be defined as logically independent of each other and of the mode of physical implementation.

An example of a data flow diagram is given in Figure 23 which is a model of the order processing requirements of ABC Ltd. The diagram is a shorthand way of identifying the data processing requirements of the system in the form of data (messages and stores) and processes. A data flow (represented by an arrow) is a collection of data items into a message. Some data flows are stored for future use; these are called data stores or files and are represented by the file symbol. A process (represented by a rectangle) is a collection of activities which transform the input data flow into an output data flow. Thus Figure 23 conveys this information about the system:

1 Customers of ABC Ltd phone in enquiries (1) about the availability of stock; these are processed (A) by looking up the stock level on the stock file (2) and giving a reply (3) to the customer.
2 Customers send orders (4) to ABC Ltd. These are checked (B) against the customer file for credit-worthiness (5) and against the stock file for availability of stock (6). If stock is available, it is allocated to the order and a note made on the stock file (7). As a result of the checking process, the orders are transformed into despatches (8) (i.e. valid orders for which stock is available) or outstanding orders (9) (i.e. valid orders for which stock was not

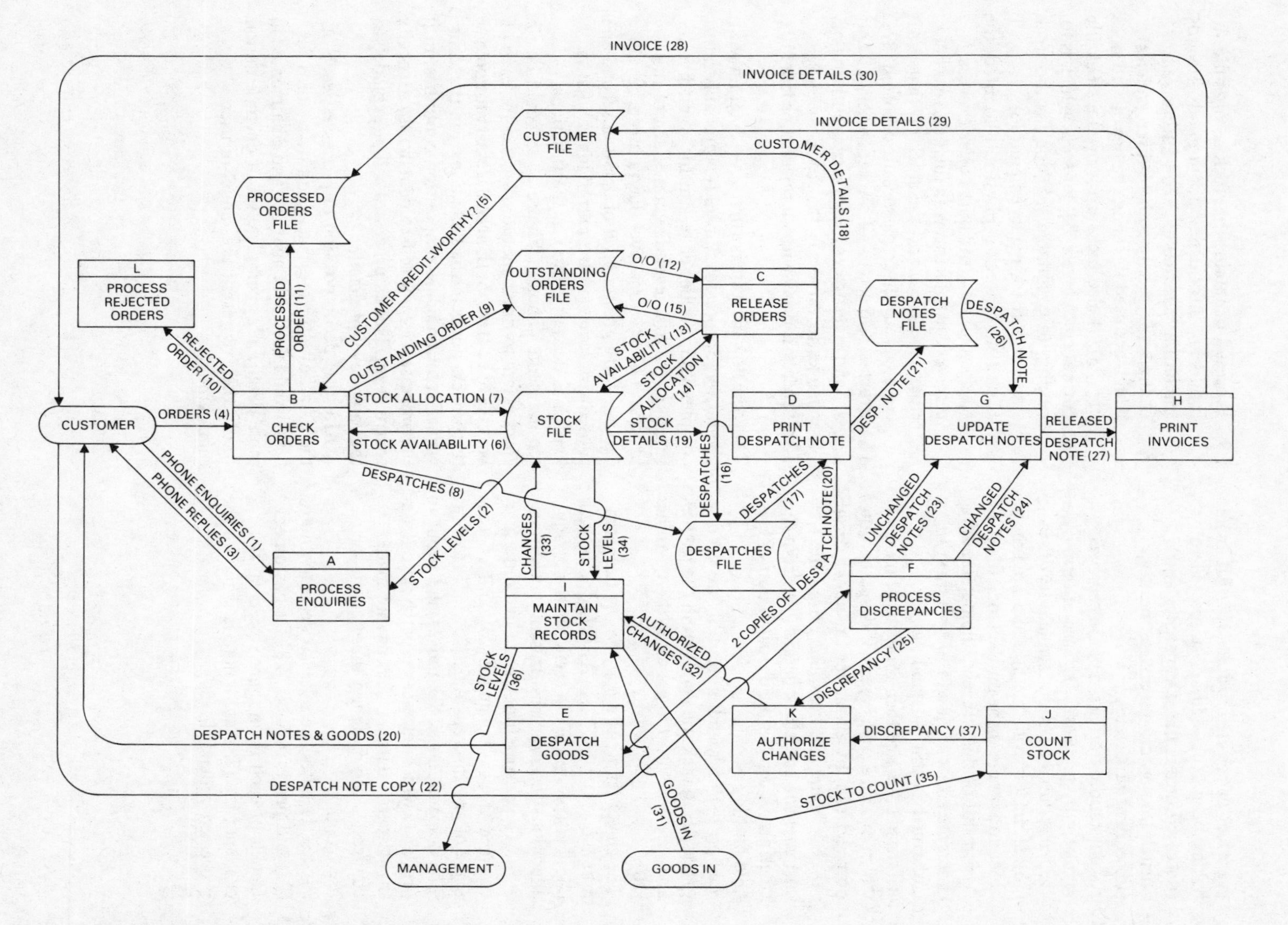

Figure 23 *Data flow diagram of order processing at ABC Ltd*

available) or rejected orders (10). All orders having been processed are stored (11) on a processed orders file. Despatches and outstanding orders are also stored for subsequent processing.

3 All outstanding orders (12) are reviewed regularly by process C (release orders) by checking the stock file for availability of stock (13). If stock is available, it is allocated to the order and a note made on the stock file (14). As a result of this process, the outstanding orders are either released as despatches (16) or confirmed as still outstanding (15).

4 All despatches are then input to process D which produces two copies of despatch notes on paper (20) by extracting customer details (18) (i.e. name, address, etc.) from the customer file and stock details (19) from the stock file. A copy of the despatch note (21) is held on file.

5 The printed despatch notes (20) are used in process E to pick the goods from the shelves and are sent with the goods to the customer (20).

6 The customer signs a copy of the despatch note and returns it (22) to ABC Ltd. It is then checked (process F) to see if there are any alterations (e.g. returns, missing quantities etc.). The despatch notes are marked as changed (24) or unchanged (23), and a note is made of any discrepancies (25).

7 The despatch notes (23 and 24) are input to process G which releases despatch notes from the despatch note file (26) having made changes as required and these go on for invoicing.

8 Process H takes the released despatch notes (27) and prints invoices (28) for customers. Details of each invoice are stored on the customer file (29) and a note is made on the processed orders file (30).

9 Process I maintains the stock records file. Its inputs are information about goods inwards (including returns) (31) and authorized changes (32) (as a result of discrepancies) which are used to change values on the stock file (33). It also picks up from the stock file current stock levels (34) for stock which is to be physically counted (35) and for the stock status report for management (36).

10 Each item of stock is physically counted (the stock-take) on a periodic basis. Process J receives a list of items with their current book stock levels (34) and the stock on that list is counted; any discrepancies are reported (37).

11 Process K receives all discrepancies (25 and 37) and investigates them prior to authorizing changes to values on the stock file (32).

12 Process L deals with all rejected orders (10).

This data flow diagram is a first-level diagram with about twelve processes shown. It is probably rather complex and it is better to aim for about five to seven on the first-level diagram. Each of the processes on the first-level diagram can be exploded to a greater level of detail in second and subsequent levels of diagram. This is what is known as decomposition of the functional requirements. Thus in Figure 23 it would be appropriate to draw another diagram for each of boxes A to L. It might then be necessary to explode the process boxes of the second level in the same way. Explosion or decomposition continues until the processes become self-evident or easy to specify or irrelevant. Not all processes will be exploded to the same level of detail.

After the decomposition of processes, each of the data flows (or messages) and data stores (or files) would need to be analysed and documented in more detail. This complete set of documentation would then represent the model of user requirements.

Exercises

It has been decided to investigate the possibility of using a computer to control the flow of traffic through the centre of a large city. The system envisaged will involve inputs from sensing devices (measuring traffic volume etc.) and computerized control of traffic lights.

5.1 Describe the kind of background information you would require if you were assigned to conduct this investigation.

5.2 Discuss the detailed information that could be gathered by observation, record searching and questionnaires.

5.3 List some of the people you would want to interview, explaining what information you would expect them to give you. Prepare a checklist of questions for one of these interviews (e.g. with the police officer responsible for accident statistics).

5.4 Design a small questionnaire for car drivers to gather information about routes travelled, times of journeys, length of journey, number of passengers and alternative transport possibilities.

6 Design

Introduction

The investigation and analysis stage of a system project will result in an outline of user requirements for the new system – possibly modelled in the form of a data flow diagram. In order to produce this, some design activity will already have taken place; so the division into stages and chapters is somewhat artificial. At some point, however, the systems analyst has to document his ideas about what the new system is going to look like. It is generally agreed nowadays that this should be done in two stages – logical system design and physical system design. The logical system is the system as it is required by the users and the physical system is the system as it is designed for a particular physical environment (including equipment, accommodation, staffing etc.). The rationale behind the split is to acknowledge that there are several physical ways of achieving the user's logical requirements and to ensure that the needs of the users are not compromised at too early a stage by the constraints of technology. After all, the computer exists to serve the users and not vice versa.

The logical system design stage involves the systems analyst in identifying the good points in the existing system, the requirements of management and the flows of information related to decision-making and overall corporate objectives, and in developing ideas about how the system could be improved (or radically altered) to meet these needs. These ideas will be documented in the form of a specification of requirements. The physical system design stage takes the specification of requirements (once it has been agreed by the various parties involved) and produces detailed specifications of inputs, outputs, files, records, codes, procedures, dialogues, forms, controls, security etc. from which programs can be written and new user procedures devised. Both of these design stages are, of course, constrained by the overall system objectives and by the resources available to the project. Thus the design must be both expedient and purposeful. The overall system objectives will vary from project to project but they are concerned with such things as:

Accuracy and reliability. The system must have an appropriate level of accuracy (this level will vary greatly from system to system).
Integration. It is important to know the system's links with other subsystems of the organization (both conceptually and in terms of the detailed aspects).
Expansibility. The system must be capable of handling all envisaged expansion during its life.
Acceptability. The system must be capable of being operated and be acceptable to the people required to operate it.
Security and control. The system must ensure confidentiality of stored data, prevent invasions of privacy, and minimize accidental or deliberate destruction or loss of information.
Cost. The system must be designed in a realistic way to meet the cost guidelines laid down by management and to ensure that the system is economically viable.

Clearly the emergent system will be designed as a compromise between these various objectives and principles.

Logical system design

The design of the logical system will take as its starting-point an outline model of user requirements as described in the last chapter. Assuming that this is in the form of a data flow diagram, the process of logical design involves successive refinement of the model until it meets all of the user requirements. The refinement normally starts with the outputs of the system and works back via files (or stored data) to inputs and procedures, taking into account the objectives and constraints that the system has to accommodate. Let us suppose for example that we are designing

Stock status report 27/6/80 Distribution – store manager
stock controller

Product no.	Product description	Qty in stock	Reorder level	Qty on order	Demand % over last year	O/S flag
187163Y	Watering can	4	6	0	+30	
187289X	Hose pipe	0	6	12	+22	*

and so on

Figure 24 *Example of a stock status report*

the system which was modelled in Figure 23. After we have discussed (over a long period and to some depth) the objectives of the system and the current methods with the user staff, we are in a position to develop ideas about the new system. We start by clarifying the various outputs of the system. Several of these have been shown on the data flow diagram in Figure 23 and each needs to be outlined and discussed with the user staff. The stock status report (36), for example, might be outlined as in Figure 24. This would be discussed with the users and amended until it was felt to be satisfactory. It would then be documented to give a formal idea of its layout, content and the medium to be used for its production. This would be influenced by the frequency and rapidity of response required; it might be a printed output or it might be displayed. Fine detail is not too important at this stage. The same process has to be followed for each of the outputs of the system. Once all of the outputs have been documented, the next step is to work out which of the data items on the outputs need to be freshly input to the process, and which can be stored or calculated within the system. Let us do this with the stock status report (Figure 24), bearing in mind that the same process would need to be done several times (for each output) in a real system.

First of all, we produce a list of data items on the output, viz

Product number
Product description
Quantity in stock
Reorder level
Quantity on order
Demand over last year (percentage increase or decrease)
O/S flag (indicator of item which is out of stock)

Immediately we can see that two of these items are produced as a result of calculation i.e. demand over last year and O/S flag. The O/S flag appears (as an asterisk) if the quantity in stock is zero; the demand over last year is calculated by comparing this year's demand for the product with last year's and producing a percentage increase or decrease. This indicates that we need to store last year's and this year's demand within the system in order to do the calculation.

Also we can see that all the data used in this report is stored data – none of it is input to the system (this is not surprising since a stock status report is basically a printout of the status of a file).

Thus we can draw a block diagram of the process which will produce the stock status report as in Figure 25. It shows that no input data is involved,

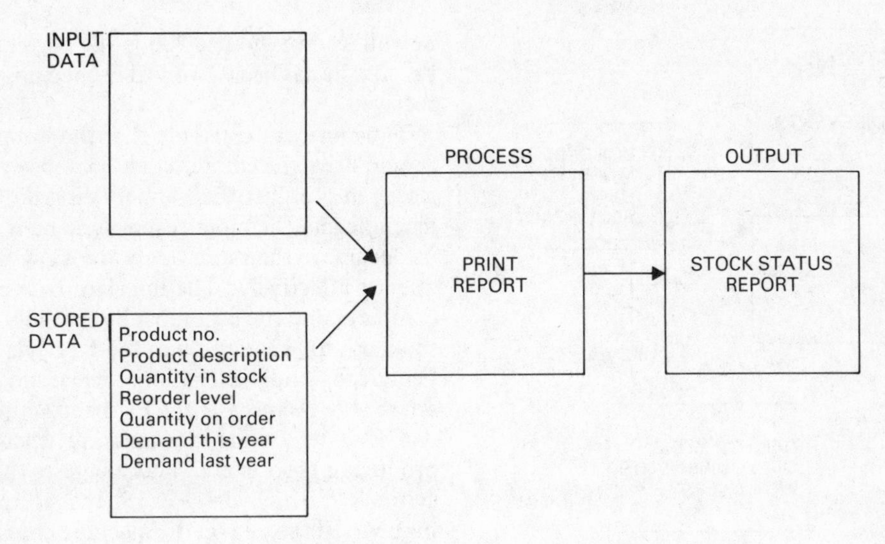

Figure 25 *Block diagram showing relationship of output to input, file, procedure*

and it lists the data required to be stored on file in order to produce the output report. The process box could be expanded into four actions:

1 Print the headings.
2 For each product on file calculate the percentage change in demand from last year.
3 For each product on file decide whether to print an asterisk to flag that the item is out of stock.
4 For each product on file print a line showing product number, description, quantity in stock, reorder level, quantity on order, demand over last year, O/S flag.

Thus we have identified in logical terms what output, stored data, input and processes are required for this small part of the system. This has to be done for all parts of all of the system with which the design team is concerned. The stored data may well be common for several of the outputs but will normally only be stored once in one location. It has to be structured in some way so that each item of stored data is related to an identifier or key via which it can be accessed. In the case of our example, the identifier of all the data items is the product code (which means that all of the data items are attributes of the entity product which is identified by the product code). One final requirement is to decide what causes the output to be produced. This can be an event (e.g. a manager requests it) or it can be the automatic result of receiving input data (e.g. a delivery note might be produced every time an order is input to a sales order processing system) or it can be a time trigger (e.g. a report on stock status might be produced at 8.00 a.m. every Monday).

All these aspects of the particular system have to be documented in the form of a user system definition (or functional specification) which can be agreed with the users. This will normally consist of the data flow diagrams together with the supporting documentation of the processes and data involved. The development of the user system definition is an iterative process illustrated by Figure 26.

Computer versus human beings

Some indication is required at the logical design stage of what the computer is going to do and what the role of the human beings will be in the system. This is a difficult aspect of design and can only really be based on judgement and experience. As a guide to the activities which are best given to the computer, emphasis can be placed on its obvious powers of storing and rapidly retrieving large volumes of data, and consistently and reliably

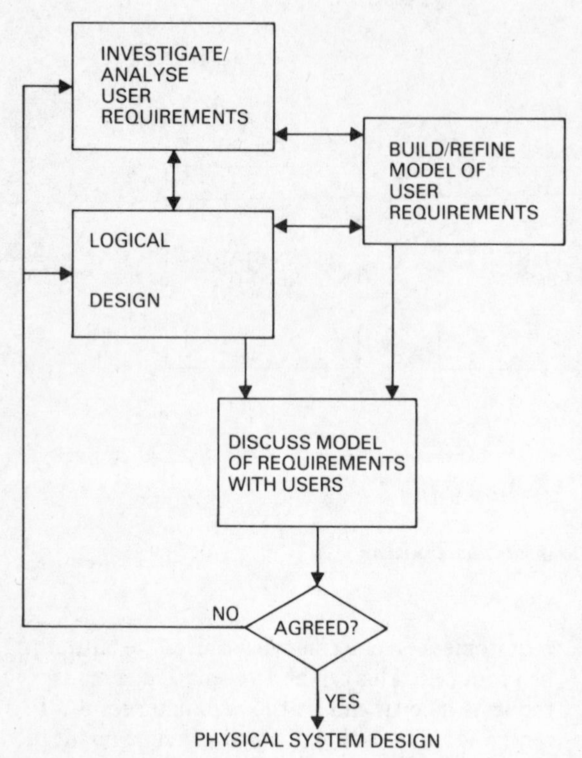

Figure 26 *Logical system design in context*

carrying out predetermined routines.

The weaknesses of computers tend to be in the areas where man is very proficient; for example, the computer is poor at communicating in a flexible way and at dealing with ambiguity. Man is good at using his experience and intuition to identify problems and opportunities, and at resolving problems of uncertainty. If a computer is told that 200 items have come into stock, it will update the file accordingly; if a man is dealing with the figures he will know whether 200 is the correct figure because he has been involved in checking the stock received.

Thus one can conclude that the human being should drive the computer and not vice versa. The closer the contact the human being has with the machine and the more responsible he is made for its accuracy, the more likely the system is to operate effectively. The interface between the computer and the user must be carefully designed so as not to separate them too far. The human being can handle exceptions, make judgements, arrive at decisions assisted by the machine, whilst it does all the routine data handling, updating files, producing reports etc. This suggests that on-line systems in which the user interacts directly with his own data stored on the machine are likely to be more successful than batch processing systems where the user tends to be more separated from the machine. However, lots of other considerations have to be taken into account (e.g. cost, security, volumes of data, hardware availability, speed and frequency of information retrieval) in deciding on the nature of the processing facilities.

The data flow diagram can be used to illustrate which activities of the new system will be computer based. A dotted line can be used to illustrate the boundary of the computer-based system in terms of the processes involved. Figure 23 is reproduced as Figure 27 with a dotted line round processes A, B, C, D, G, H, and I; these are the processes which will be computerized. Some of these processes have the dotted line running through them (A and B); this indicates that the task will involve direct man–machine interface for decision-making. These are the processes which are likely to require on-line interaction; and others can be entirely computer based.

Physical system design

When the logical system has been defined to the satisfaction of the user, work can begin on detailed physical system design, i.e. detailed specification of the way the system will operate in a specific environment, using specific equipment and specific people. There are lots of tasks involved in physical system design, covering all aspects of computer-based systems. For the sake of simplicity they will be broken down here into a series of separate tasks, though in practice they

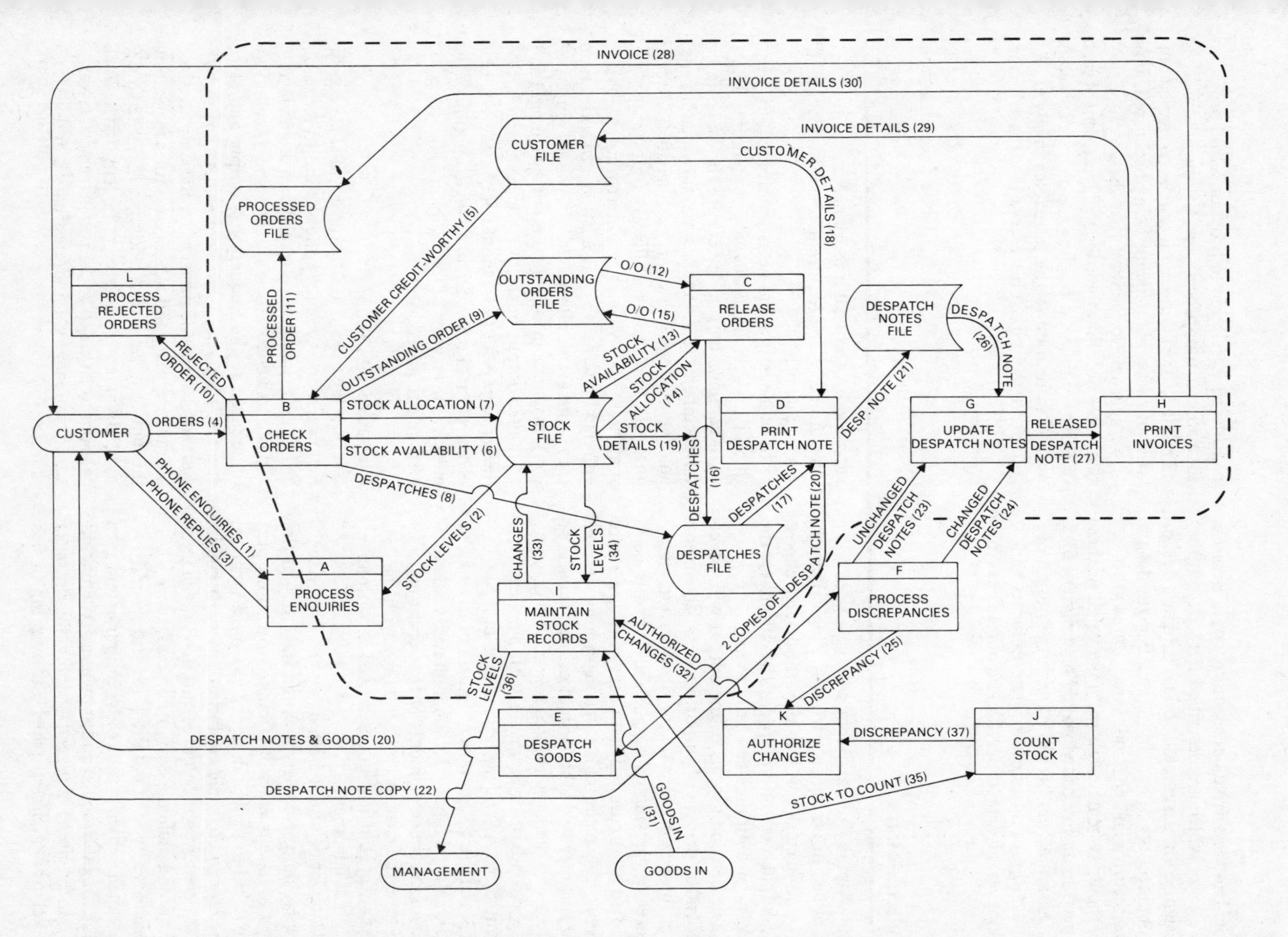

Figure 27 *Data flow diagram of order processing at ABC Ltd, highlighting computerized processes*

would not be carried out separately or in isolation. The tasks which are described here are output design, input design, file design, program design, user procedure design, forms and dialogue design, and security and controls. Any approach to computer system design will involve all of these aspects in an integrated way. The design of files, for example, will be tied in with the design of input; input must closely relate to forms and dialogues; all data (input, file and output) has to be considered in relation to programs, and programs in relation to user procedures. In other words, they all relate to one another and cannot be carried out without these relationships being optimized. It is also important to mention that the scope of this book does not allow for a detailed treatment of each of these complex tasks; they are only discussed in outline in relation to the principles involved. A fuller treatment must be reserved for other books.

Output design

The outputs from a system will have been identified in outline as part of the logical system design. The task now is to fill in the details, choose the output medium and specify the outputs for the benefit of both user and programmer.

The detailed aspects of the output will largely depend on the type of output. External documents (i.e. those going outside the organization and therefore carrying its image) will normally be printed on preprinted stationery with great care given to the quality of layout and printing. Internal outputs may be in the form of printed reports or interactive displays. Printed reports will generally be on standard listing paper with headings etc. printed by the computer; neatness of layout is important (e.g. columns of figures neatly spaced, subtotals and totals at frequent intervals, adequate spacing between elements of the report etc.) because the user has to feel happy with the document, but it is not as crucial as with an external document.

Interactive outputs are displayed on a visual display unit or a teletypewriter and, though they tend to have a short life (unlike printed output), they need to be well designed to aid the terminal operator and to avoid confusion. (More discussion of the content and layout of outputs is given under forms and dialogue design). A final, more specialized type of output is the 'turnround' document which is printed by computer and then has extra information added by users (via machine or hand) before being read again by the computer. Clearly the quality of paper and printing in this situation is of crucial importance and will be determined by the requirements of the document reading device.

Once the output has been determined, the next thing is to determine the output medium. The most common output devices are the line printer for documents (e.g. invoices, payslips, statements) and reports (e.g. sales analysis, production scrap analysis), and the visual display unit (or, sometimes, typewriter) for interrogations, data entry and interactive conversations. In addition to these, the graph plotter (for drawings, graphs etc.), magnetic media (for temporary storage) and computer output microfilm (for archiving large files or large print runs) should be considered. The choice of medium will depend on the nature of the output, the location of the recipients, the speed of response required, and the cost.

Output specification

It is now possible to look at the detailed aspects of the output and to specify the content in terms of headings, fields, totals etc., the format and layout, and the sequences. In addition, some thought has to be given to the frequency of output and the speed of response. Each item of data has to be defined in terms of position on the output, size, character types, valid ranges etc. Rules for editing the data (e.g. removing leading zeros, inserting £ sign, alignment etc.) have to be laid down. Each page or screen has to be defined in terms of layout, size, frequency, volume, destination and action required.

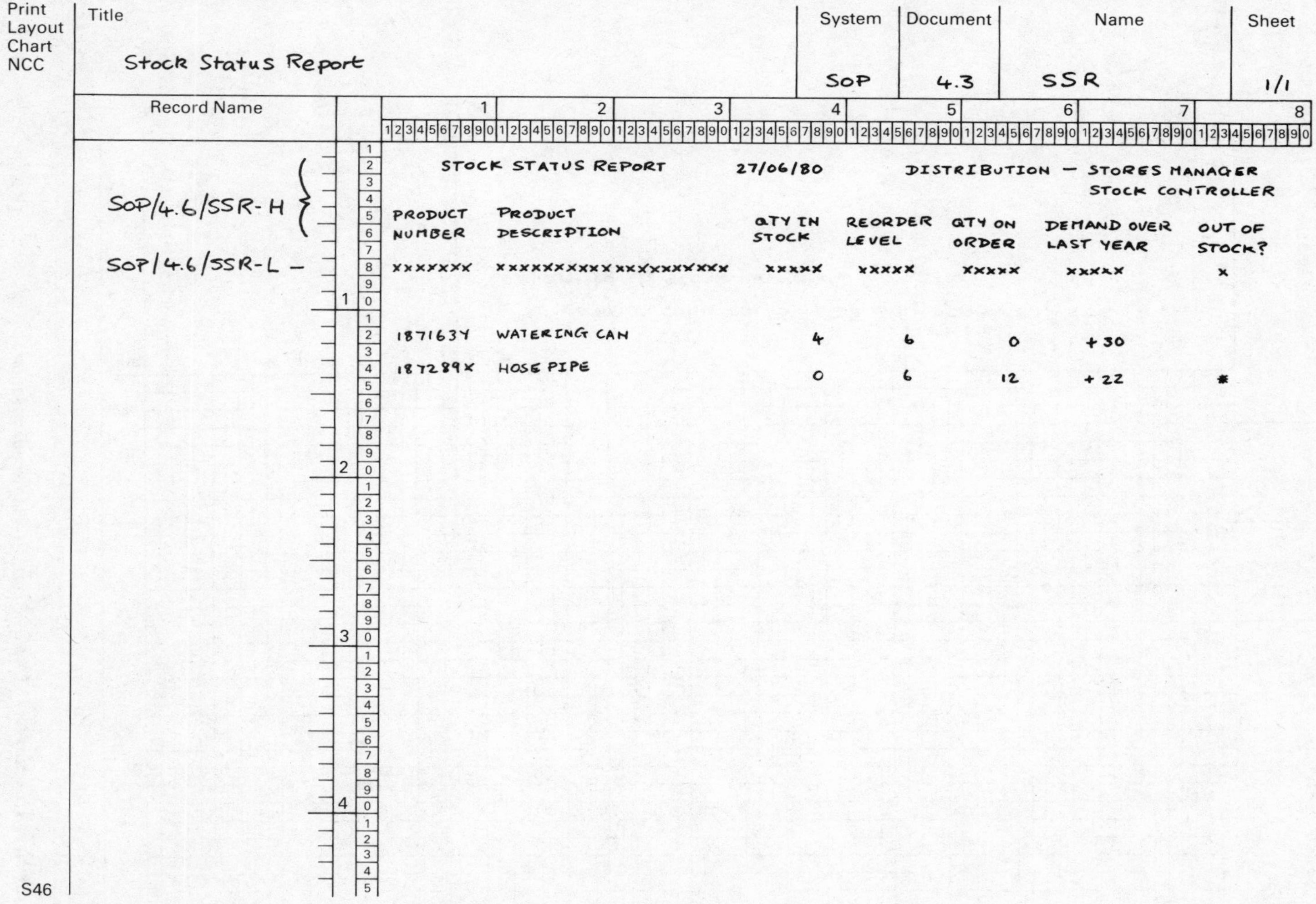

Print Layout Chart NCC

Title	System	Document	Name	Sheet
Stock Status Report	SOP	4.3	SSR	1/1

Record Name	Line	Product number	Product description	Qty in stock	Reorder level	Qty on order	Demand over last year	Out of stock?
SOP/4.6/SSR-H	2		STOCK STATUS REPORT	27/06/80		DISTRIBUTION — STORES MANAGER		
	3						STOCK CONTROLLER	
	5–6	PRODUCT NUMBER	PRODUCT DESCRIPTION	QTY IN STOCK	REORDER LEVEL	QTY ON ORDER	DEMAND OVER LAST YEAR	OUT OF STOCK?
SOP/4.6/SSR-L	8	XXXXXXX	XXXXXXXXXXXXXXXXXXXX	XXXXX	XXXXX	XXXXX	XXXXX	X
	12	187163Y	WATERING CAN	4	6	0	+30	
	14	187289X	HOSE PIPE	0	6	12	+22	*

S46

Figure 28 *Example of a completed S46 Print Layout Chart, showing the specification of the output report, stock status report (Figure 24)*
The print layout chart is usually 160 columns wide and 66 lines deep. It has boxes to indicate each print position on a sheet of standard listing paper

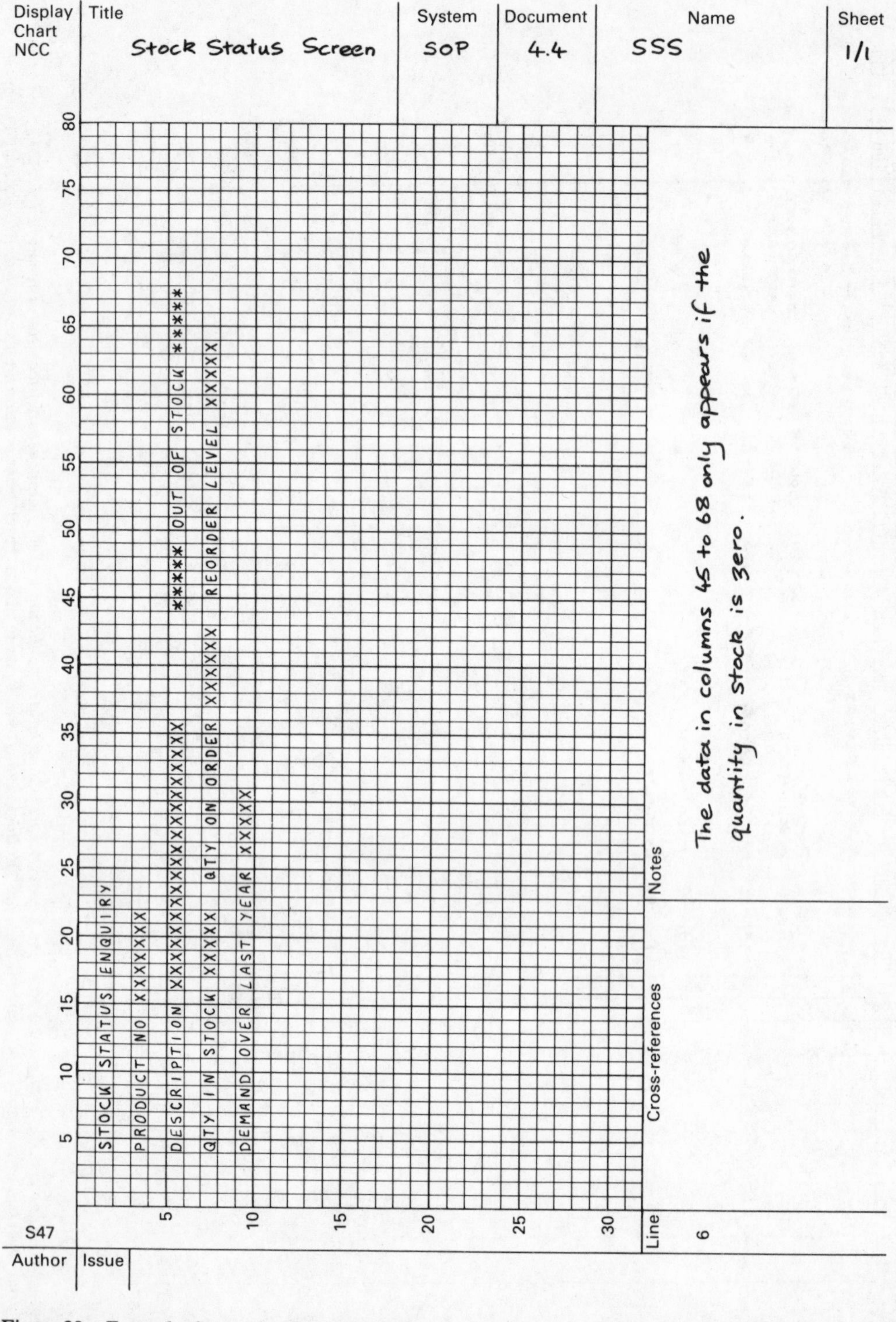

Figure 29 *Example of a completed S47 Display Chart, showing an interrogation about the status of a particular product's stock level*

Computer Document Specification NCC	Document title	System	Document	Name	Sheet
	Stock Status Report	SOP	4.3	SSR-CDS	1/1

Stationery ref.	Width	Depth	Number of parts	Blank/~~preprinted~~
Standard listing paper	—	—	2	

	Average	Maximum	Growth rate
Pages	1000	1200	3 % per annum
Lines per page	30	30	or determining factor

Page and line spacing/stepping: Double line spacing after headings

Ribbon type	Ribbon life	Printer speed	Lay-out chart ref.	Control loop Ref.
Standard	—	—	4.3/SSR	Standard

OUTPUT ONLY

Part No	Trim/Burst	Distribution	Line	Channel
1		Stores Manager		
2		Stock Controller		

MACHINE READABLE ONLY

Clear area – distance from edges

Reading method and font

Source

Level	Record Name	Size	Unit	Format	Occurrence
GA	STOCK STATUS REPORT	5 – 34	L	F	500 – 1200
B	SSR-H	4	L	F	1
B	SSR-L	1	L	F	1–30

S43

Figure 30 *Example of a completed S43 Computer Document Specification, showing the specification of the stock status report in Figure 28*

The presentation of output is normally in the form of a layout chart which can be used both for the user to gain an idea of the format (even better would be a mock-up on actual stationery or screen of the output) and for the programmer to show where particular headings and fields should be printed. A layout chart for a printed output (i.e. a print layout chart) or a displayed output (i.e. display layout chart) would consist of graph type paper showing the lines and columns of a page or screen. The NCC Print Layout Chart (S46) and Display Chart (S47) would be used for this purpose.

Figure 28 gives an example of the use of a print layout chart to define the output report which was shown in Figure 24. On this the headings are shown in their required form and the variable information is shown as a number of *x*s which indicate the maximum size of the field to be printed (the editing of the field will need to be defined on a separate sheet). This layout indicates to the programmer how each of the print lines on a page is to be laid out. Figure 29 gives an example of a display layout chart for an interrogation similar to the report in Figure 24 where a question has been asked about the status of a particular product. In each case the detailed specification of each data item appearing on the print layout or the screen would be specified on a Record Specification (one for each line) of which there is an example in Figure 34.

The structure of a printed report and its physical characteristics (e.g. paper type, size, number of parts) need to be defined also. The NCC form for doing this is the S43 Computer Document Specification; an example of this is shown in Figure 30 where the report shown in Figure 28 is specified.

Input design

Input design is a major part of physical system design because the decisions involved will have a major effect on the cost of the system and its acceptability to the users. Providing input data for the computer in a relatively speedy and error free way is a major system problem. Before detailed input design begins, some idea will already have been gained about the nature of the input data; the task now is to determine detailed aspects, to choose the input medium and to specify the input data for the programmers.

Data capture methods

There are four main methods of capturing input data – source document conversion, by-product data capture, direct data capture, and on-line data entry.

Source document conversion

This is the process involved when input documents (e.g. customer orders, job cards, time sheets, file amendments etc.) are batched together and then converted (normally by punching) into a computer-acceptable medium. The most common methods used are punched cards, punched paper tape, and keying to magnetic media (tape, cassette, cartridge, diskette or disk). The data capture devices have keyboards and the operator types the input data via the keyboard on to the appropriate medium which is then loaded on to a computer input device to be read into the computer.

By-product data capture

This is the name given to the method whereby input data is captured on a device which has some other main function. A cash register, for example, which is being used to record customer purchases and calculate totals, can be fitted with a paper tape punch or a cassette recorder or an optical character printer. The computer input medium is removed from the machine at the end of a period and loaded on to a computer input device.

Direct data capture

This is the process of capturing input data directly without a conversion stage. Marks filled in on questionnaires which can be read directly, bills produced by a credit card inprinter, and tags

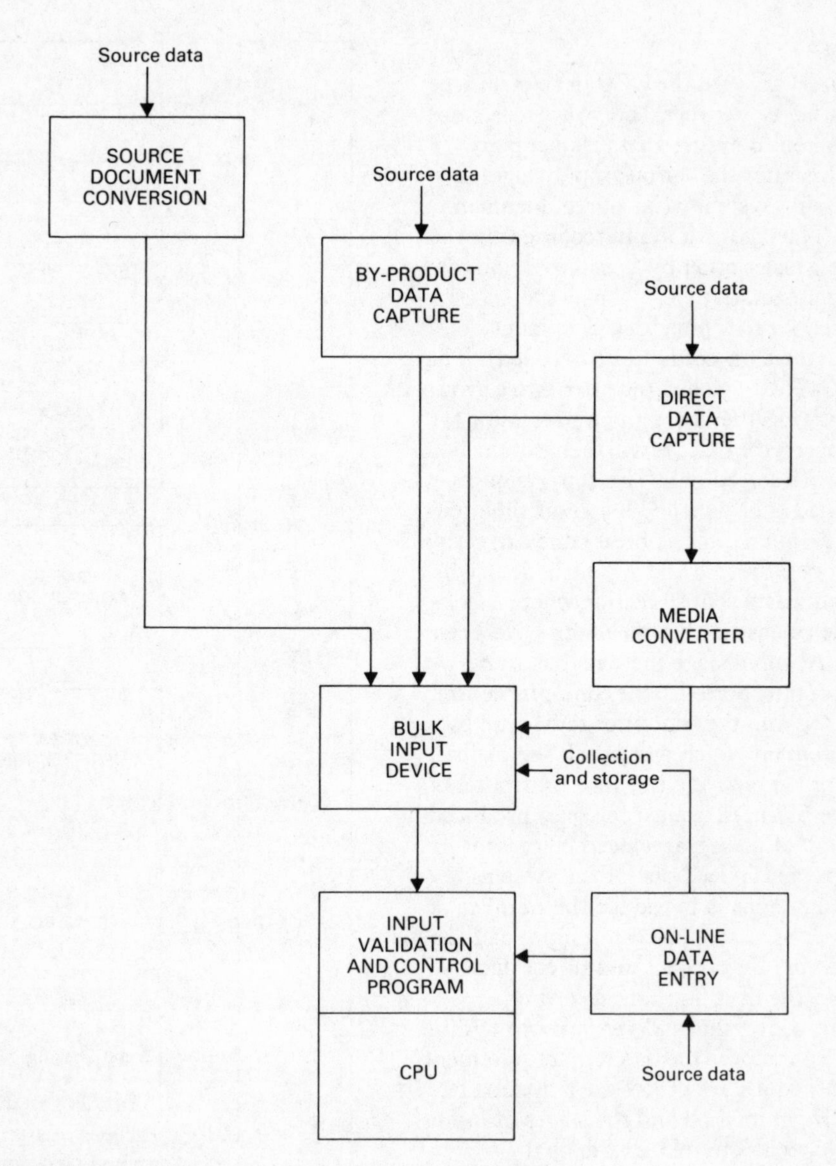

Figure 31 *Relative closeness to the machine of the data capture methods*

attached to clothing and collected at the point of sale are examples of direct input media. Sometimes the data (e.g. from tags) is converted to a faster medium before input.

On-line data entry

This is the activity involved when data is entered into the computer under program control, one transaction at a time. Shop floor data collection devices, light pens, visual display units, teletypewriters and analogue to digital converters are the main types of on-line data entry device. Often these devices can be used off-line to collect data which is then fed into the computer in bulk. Figure 31 illustrates the relative closeness to the computer of the various methods; the closer the method, the faster the input data reaches the computer, but it is usually more costly.

Input stages

Input data must arrive at the CPU in an error free condition as far as possible, and so various steps have to be taken to ensure that this happens. Figure 32 illustrates the various input stages in a batch processing system with source document conversion. The data is initially recorded by the originator (e.g. sales order by a customer, job card by a shop floor operative etc.). It may then need to be transcribed on to a punching document (especially if the data needs to be encoded). The transactions will be batched together into a group which will stay together throughout processing for control purposes, the batches will then be punched and any queries sorted out. The conversion may well be verified (i.e. punched a second time to check that the punching has been done correctly) and any errors corrected and repunched. It will then pass to the data control section where checks will be made to ensure that no errors have been introduced. At some stage the data has to be transmitted or transported to the computer centre. Finally it is fed into the computer to be read by a validation program which will check the logical correctness of the data and produce control totals which can be balanced against the ones produced earlier. Not all of these stages need be present in every system, and indeed one of the systems analyst's design aims is to reduce the number of stages in order to lessen the cost of input and the chances of error. By-product and direct data capture avoid the need for data recording, transcription, conversion and verification; on-line data entry also removes the batching requirement because transactions are processed individually.

Once the input method and the stages of input activity have been determined, the final requirement is to specify the input method. This will involve the design of input procedures (both computer and clerical aspects), forms and possibly dialogues, and specification of input records which relate directly to the input medium (e.g. fixed length records of maximum 80 characters on an 80 column punched card). These topics are discussed at greater length later in this chapter.

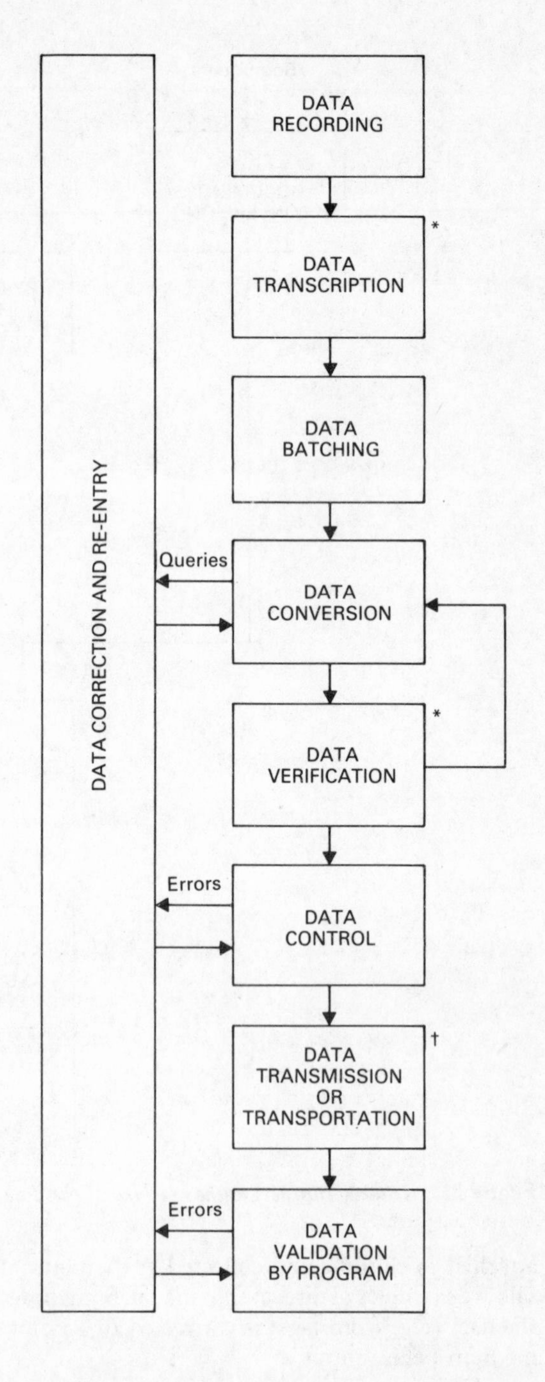

Figure 32 *Input stages in a batch processing system with source document conversion*
Notes: * These are optional stages.
† The point at which data transmission or transportation takes place will vary from system to system.

File design

The design of output and input is relatively straightforward but absolutely crucial because they are the points at which human beings come into contact with the computer system. The design of files is more complex because it will directly affect the efficiency of the system and it is integrally linked to program design.

Types of file

There are various types of file used within computer systems. The main one is the *master file* which tends to contain the permanent data within the system against which transactions are processed. For example, in a payroll system one of the master files would be the employees' file against which transactions such as time sheets would be processed in order to produce payslips, and in a library system two of the master files would be the book file and the borrower file which would be linked to establish who had borrowed which book. Master files normally contain reference data (i.e. static information describing the entities under consideration), which is processed by amending (i.e. irregular changes in the form of insertions of new records, deletion of existing records, and alterations to the existing records), and dynamic data (i.e. data which is constantly changing), which is processed by updating (i.e. changing the values of particular fields). For example, a vehicle hire file would consist of items of data describing the vehicle (e.g. its type, size, colour, age, hire charge etc.) which will rarely change (i.e. reference data) and items of data describing the hire of the vehicle (e.g. the name of the hirer, the period of the hire, special aspects of the hire etc.) which will frequently change, perhaps every day (i.e. dynamic data).

Some other types of file tend to be known as transaction files. These include input files which hold data coming into the system, output files which hold data which is about to go out of the system, and transfer files which hold data temporarily between programs either within the same system or between systems. Finally there are dump or work files which are used to hold data for security and recovery purposes or for temporary storage whilst other processing is being carried out.

File structure

A file will often have a hierarchical structure; it may consist of subfiles which will have records which will be made up of items of data. (In the most recent approaches to file design, facilitated by more advanced software, this hierarchical structure has often been superseded by other structures in which records are linked together in an integrated way without the superimposition of files. This is known as the data base approach to file design but there is insufficient space to discuss it here.) A file can consist of any number of records and a record of any number of data items.

The records and the data items can be fixed or variable in length, but the greater the variability the more difficult the programming becomes. From a physical storage viewpoint, records tend to be grouped together into what are known as blocks, and the block is the physical data which is read into main storage from the backing storage device. So when a program reads some data, it reads a block and this may consist of one or more records.

File organization

The way that records are arranged within a file on a particular device is known as file organization. On magnetic tape, because it is a serial medium, records have to be read and written one after the other; this is the physical sequence of reading and writing. They can, however, be organized logically in two ways – either randomly or sequentially. In random organization the identifier (key) of each of the records which make up the file has no relationship with the previous or next record in the file; in sequential organization, the records are in sequence (ascending or descending) of their identifiers (keys) and so each record has a direct logical relationship with the next one. In order to achieve this sequence, files on magnetic tape have to be sorted into the required sequence by program.

On magnetic disc it is possible to read and write

Computer File Specification	File Description: CUSTOMER FILE	System: SOP	Document: 4.4	Name: CUST FILE	Sheet: 1/1

NCC

File type: Input ☐ Output ☐ Master ☑ Transfer ☐

File organisation: INDEXED SEQUENTIAL

Storage medium: Meg. tape ☐ Disc ☑ ☐ Single ☑ Multiple ☐

Retention period	Number of generations	Number of copies
		2

Recovery procedure: Daily dumping

Keys: Customer account number

Labels: Standard

Level	Record name/ref.	Size	Unit	Format	Occurrence
A	4.7/CUSTREC	240	B	F	10,000

Block/batch size

Actual, for fixed length	Maximum, for variable length
240	—

File size

Average	Medium
2,400,000	2,640,000

Unit of storage: Records ☐ Words ☐ Bytes ☑ Characters ☐ Cards ☐

Number of blocks

Average	Maximum
10,000	11,000

Growth rate: 2 % per annum or determining factor

MAG. TAPE ONLY	Tracks: 7 ☐ 9 ☐	Recording density	Speed	Length

DIRECT ACCESS ONLY

Addressing/accessing method	Packing density	Frequency/condition of re-organisation
Index	90 %	Monthly

Level	Type of overflow	Size of overflow areas

Notes

S42

Figure 33 *Example of a completed S42 File Specification, showing details of physical aspects of a customer file, which has only one record type*

Record Specification — NCC — S44

Record description	System	Document	Name	
Customer records	SOP	4.7	CUSTREC	1/1

Medium	Record length	Record format	Record size	Words / Characters / Bytes	File specification refs.	Lay out chart ref.
DISK	Fixed ☑ Variable ☐	Fixed ☑ Variable ☐	240	Words ☐ Characters ☐ Bytes ☑	SOP/4.4/CUSTFILE	

Ref.	Position From	Position To	Level	Name: In system design	Name: In program	Data Type	Size	Alignment	Picture	Occurrence	Value Range
1	01	06	03	Customer account number			6		9(6)		100000-499999
2	07	126	03	Customer name and address							
3	07	36	05	Name			30		X(30)		
4	37	126	05	Address line			30		X(30)	3	
5	127	216	03	Delivery address							
6	127	216	05	Address line			30		X(30)	3	
7	217	218	03	Discount category			2		X(2)		01-14
8	219	222	03	Salesman number			4		9(4)		1001-5999
9	223	225	03	Region code			3		9(3)		101-599
10	226	228	03	Main product category			3		9(3)		001-999
11	229	234	03	Credit limit			6		9(6)		000001-999999
12	235	240	03	Current balance			6		9(6)		000001-999999

Figure 34 *Example of a completed S44 Record Specification, showing the content and structure of the customer records which make up the customer file*

records either serially (i.e. one after the other) or randomly (i.e. wherever on the device it is required for the next record to be read or written). Thus records can be organized logically in three ways – serially, sequentially and randomly. Serial organization is when each record is copied on to or copied from the device in the next physical location – it tends to be used for transaction and dump files. Sequential organization is when it is possible to find a relationship between the records on the device via their keys (i.e. having found a record with the key 10763, it is possible sequentially to find record 10764). Random organization is when each record on the device is located with no regard for the other records nearby – its position is found by some direct addressing method. Sequential and random organizations are used mainly for master files; where transactions are grouped for processing, sequential is favoured, and where the next transaction bears no logical close relationship to the last, random is used. Both sequential and random methods of organization have the problems of catering for expansion in file size. This is catered for in the technique known as overflow, whereby overflow areas are allocated on a device to cater for the expanded or new records.

Access methods

Records are thus organized within files, but means are needed to access the records stored in these ways. The access method for serially organized files and for sequentially organized files on tape is serial reading and writing (i.e. each record is read in physical sequence from the device). For sequential and random files on direct access devices, other methods of access are required and the most common are indexes, address generation, and chaining.

Indexing involves the creation of an index to a set of records which is accessed to find the address of the required record. The key and the address of each record are held in the index and the address is found by searching for the key. If the index is large it may itself be indexed into ranges, and there are various different techniques available for indexing.

Address generation involves the use of a key transformation algorithm to produce the required address of the record. Here the key of the record has some mathematical calculation applied to it (e.g. division, truncation, folding, squaring) to derive the address; clearly careful design of the algorithm is required to ensure an even spread of keys to addresses and to avoid synonyms.

Chaining involves the use of pointers in records to identify the address of the next record in a required sequence. The pointer is a data item within the record which normally is the key of the next or prior record in the chain. More than one pointer can be used to give links to the next, prior or end of chain record. This access method is particularly useful for the data base approach where linking of records is an essential requirement.

Other more advanced methods of file organization include file inversion, where essentially records are identified and accessed by fields other than the identifier field. Thus in an inverted file it would be possible easily to bring together or to process all the records of employees who had been on a particular training course; with other methods of organization this would only be possible if the training course was the key of the employee record. Thus file inversion is used mainly in interrogation.

Design issues

File design consists of making choices among these various possibilities and then specifying the files and records required by the system. The analyst rarely has a free choice because any given system has to fit the available hardware and software and interface with other existing computer-based systems. The choice of file medium will largely be governed by the purpose of the file, the response required to enquiries, the size of the file and the file activity. On the whole it is true to say that a file which has to be available on-line during the day, has to offer a response of less than 5 minutes, is not too large and is frequently used will have to be stored on a direct access medium. A file which is processed once per day and is quite large will tend to be stored on a serial access medium. Files in between these will be stored on one or the other depending on the particular circumstances of the installation.

The method of file organization chosen will be affected by file volatility, file size, and hit rate. File volatility is the extent to which the records on the file change over time; a highly volatile file would cause enormous problems both for overflow and for randomly organized files. File size is important because it affects the amount of storage committed (especially in an on-line situation where the file needs to be available during the day for fairly fast response situations). Hit rate is the number of records accessed in a typical processing run, expressed as a percentage. The higher the hit rate, the more likely that batch processing of a sequentially organized file is needed; the lower the hit rate, the more appropriate becomes direct processing of a randomly organized file.

This section has skimmed over the surface of file design. Very complex issues are at stake when files are designed and many of them can be assessed quantitatively. Block sizes, response times, index sizes, hit rates, key transformation algorithms etc. are examples of file design problems with a quantitative aspect, which is tackled in other books.

Specification of files and records

Once the records and files have been designed they need to be specified. The file specification needs to record the structure of the file in terms of records, and various aspects of the physical implementation such as organization method, medium, security procedures, volume of data, block size, accessing method, overflow arrangements etc. The NCC File Specification (S42) is used for this purpose; it is illustrated in Figure 33. The record specification needs to show all the data items of the record with their structure and format. The NCC Record Specification (S44), which is shown in Figure 34, follows the COBOL convention in its method of recording the structure and format (picture) of the record. The record specification would be used to specify any message of a system which is input to, or output from, the computer, whether on a file, a printed report or a screen.

Program design

The systems analyst's task with regard to programs is to provide to the programmer(s) a sufficiently detailed statement of what the computer procedures are intended to achieve so that the programmer(s) can produce the programs. The extent to which the systems analyst defines each program depends on the standards used in the installation. In some computer departments a senior programmer will be included in the project team to assist with the design of programs, and so programs may be individually specified even to the extent of outline logic. In other situations the systems analyst will define the overall requirements in terms of the outputs required and the files and inputs to be used, and the programmers will work out the combination of programs required. In this case, the analyst will provide a program suite specification to the programmers rather than a program specification. Whichever is provided, it will act as a kind of contractual document which defines the job that the programmer has to perform for handing over to the systems analyst.

Regardless of the extent to which the systems analyst specifies individual programs, he has to have an understanding of program design because his decisions on file organization and access methods will greatly affect the programs. If, for example, a file is to be stored on magnetic tape, then it can only be processed serially; this determines the nature of the processing program.

The main tools used by the systems analyst in defining procedures are flow charts, structured English and decision tables, which are described in detail in Chapter 10. In program design, an overall program suite will be depicted by a computer run chart and program logic by structured English, procedure charts or decision tables. None of these is intended to determine the structure of a program or its detailed logic (which are the responsibility of the programmer); rather they are used by the systems analyst to attempt to clarify the procedures which he requires and which tend to be confusing when presented simply in narrative form.

Program types

There are four main types of program in any given computer system, though each of these types may be broken down into a series of programs or combined into one large program in practice. These four types are the input or validation program, the sort program, the file processing program and the output or print program. They are illustrated together in the computer run chart in Figure 35.

Input or validation program

This typically reads the input data in the form of a

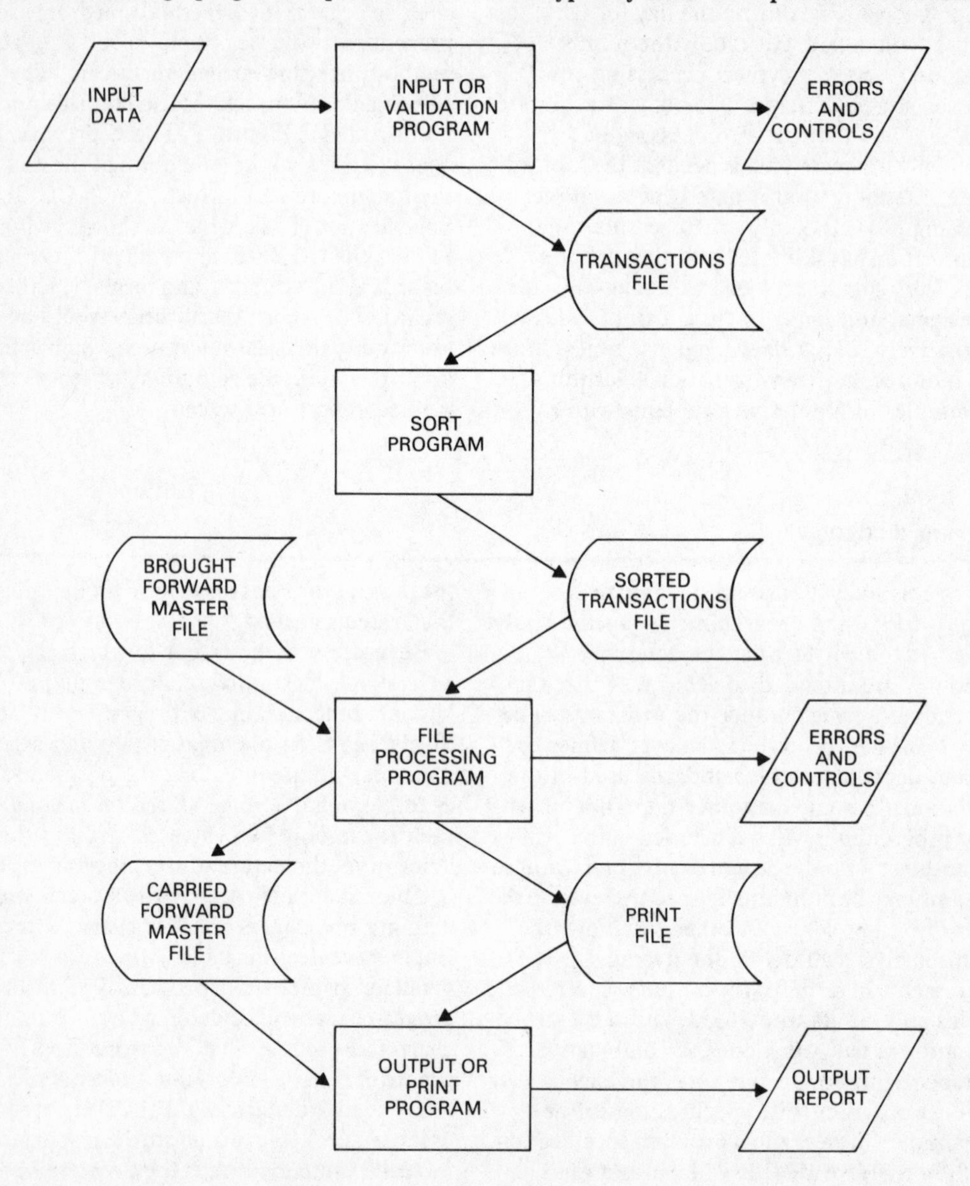

Figure 35 *Computer run chart for a typical simple batch processing system where the master file is on a serial medium*

group of transactions and loads them on to a backing storage medium, having checked the accuracy of the data as far as possible. If there are any errors they are printed out for subsequent correction and re-entry. Control totals are produced of the number of records handled and their total value for comparison to manually produced control totals. Thus if the input data to a student records system consisted of a magnetic tape record for each student giving student number, name, address, course, fee paying authority and academic qualifications, the validation program would read each record at a time from the magnetic tape and check that each of the fields of the record was valid (i.e. correct length, format, etc.). The valid records would be written on to a backing storage device, and the invalid ones would be printed out on a line printer.

The valid transactions are going to be used to update a master file, and our assumption here is that the master file is organized on magnetic tape and so the transactions have to be sorted into the sequence of the master file before they can be processed against the master file. (If the master file had been stored on a direct access medium, and could be updated randomly, it would not be necessary to sort the transactions. In this case the computer run chart might look like Figure 36.)

Sort programs
These tend to be provided as utilities in manufacturer's software, and so appear in a variety of forms, sometimes as an option within a language, sometimes as stand alone programs. The input is usually a file on magnetic media and the output is a sorted file on magnetic media. The

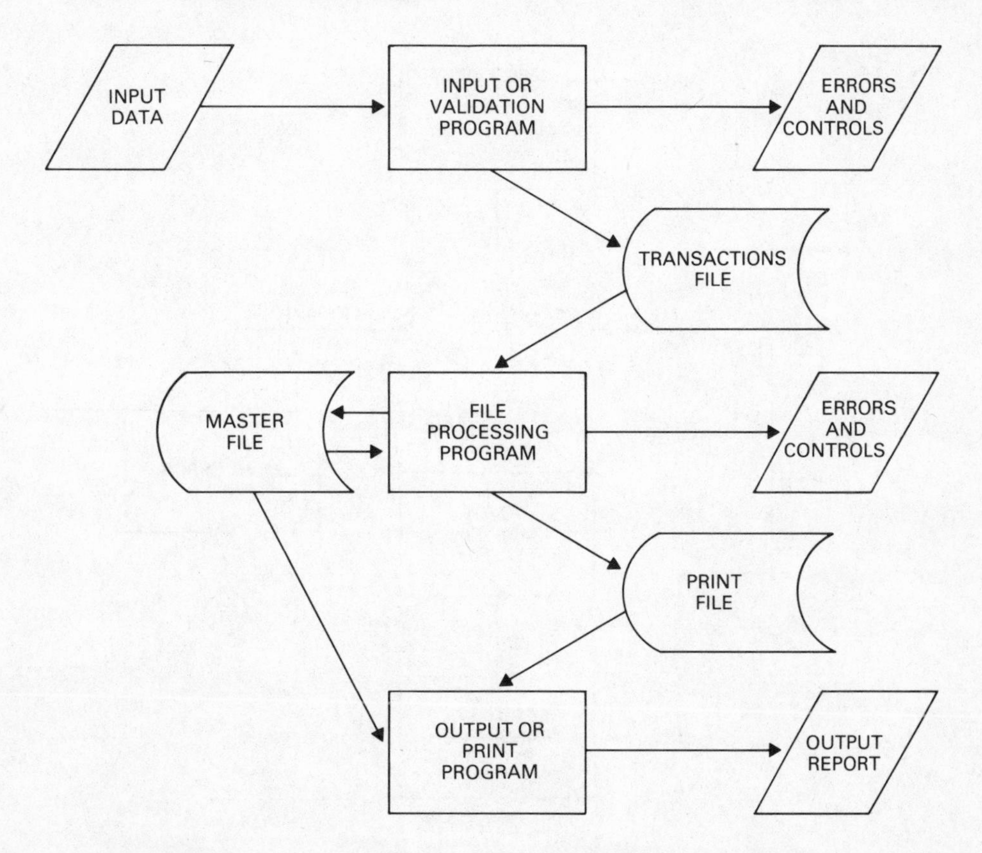

Figure 36 *Computer run chart for a typical simple batch processing system where the master file is to be accessed randomly and is stored on a direct access medium*

program is controlled by parameters which are read in on cards or paper tape and which have two main roles. The first is to identify the field of the records which is going to be used to sort into the required sequence (i.e. the sort key). If the sort was intended to put student records into student number sequence, then the sort key would be student number; if student name sequence was required, then student name would be the key. The sort when executed would put the records on the output file in the sequence of the key, using the collating code of the particular machine.

The second role of the parameters is to allow a programmer to add own coding to the sort. In first-pass own coding, each record is presented to the program before being passed to the sort; in last-pass own coding, the records are passed to the program after the sort and before being written to the output file. Other similar facilities are available in manufacturer's software, and the aim is primarily to allow the programmer to edit fields on the file or to make relatively small additions to the procedure to make the sort or the suite of programs more efficient.

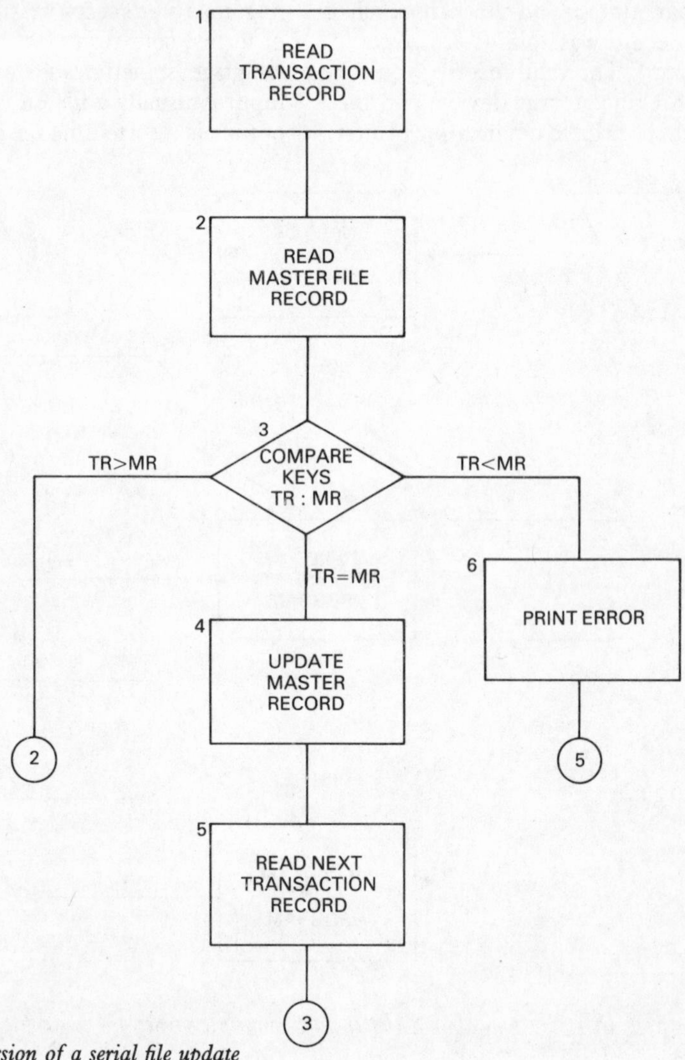

Figure 37 *Simplified version of a serial file update*

File processing programs
These are the most complex type of program to define and write – and the complexity increases with the number of files against which transactions are being matched. The program typically reads a transaction from the sorted transaction file and then reads records from the master file until the master file record with the same key as the transaction is found and the master file record can be updated or amended. (In the case of direct access processing the master file record will be accessed directly, given the key of the transaction record.) A computer procedure flow chart in Figure 37 shows a simplified version of a serial file update. A transaction record is read, followed by a master file record; the keys of the records are matched and, if they are equal, the master record is updated, the next transaction read in (there may be more than one transaction for the master) and the key comparison performed again. If the transaction record key is greater than the master record key, then the next master record is read in until equality is found. If the transaction record key is less than the master record key then the transaction record has no corresponding master file and so there is an error which must be printed before the next transaction record is read. This is a very simple procedure involving just one file and simple updating. If the file was being amended as well (i.e. new records inserted, records deleted, and alterations to reference data) then the procedure would be more complex. The file processing program will normally be required to print errors (which are usually the result of mismatches) and control totals (showing the number of records processed to be balanced against the controls produced by the validation program). In addition it may be required to produce an output file (as well as an updated (carried forward) version of the master file) for subsequent processing.

Print program
The final program type is the print program, which is required to output the results of the updating/file processing procedures. Sometimes the print program will print from the carried forward master file, sometimes from a print file. Print programs are often provided as utilities by manufacturers. The basic task of the print program is to extract appropriate data from the files, to set up lines of print for the line printer and to control the movement of the printer. The systems analyst has to define printouts carefully so that printing efficiency and speed are optimized.

In an interactive system, this division into program types is less meaningful. The terminal operator will communicate with segments of programs called in to do the various tasks required at a particular point in time, and will tend to communicate with a control program which organizes the processing. Thus the computer procedure may look like Figure 38.

Design issues
The considerations that affect the design of program suites and programs are many and complex and they can only be touched on lightly here. First of all, the nature of the programs will be greatly determined by the hardware available (especially the backing storage), which will affect the method of file organization, and the amount of main storage, which will influence the size of the programs. The number of peripherals required, the size of the operating system, the blocking factors adopted for files, the availability of buffers etc. will all affect the amount of storage space available for programs. Second, programs are obviously constrained by software availability; the operating system (and especially the use of multiprogramming), the utilities (including application packages), and the language compilers will have a direct influence on the nature of programs. Third, the design of the programs, the approaches to file (in particular), output and input design, the requirements for security and controls, the amount of computation or printing, the merging of updating and amendment routines into single programs, and the degree of flexibility required are all factors of importance. The optimum solution for any system is one program with the minimum of loading of peripherals etc.; this is rarely achievable, but the aim is to keep to a minimum the segmentation of procedures.

A lot of the decisions which have been hinted at in the previous paragraph are decisions for the

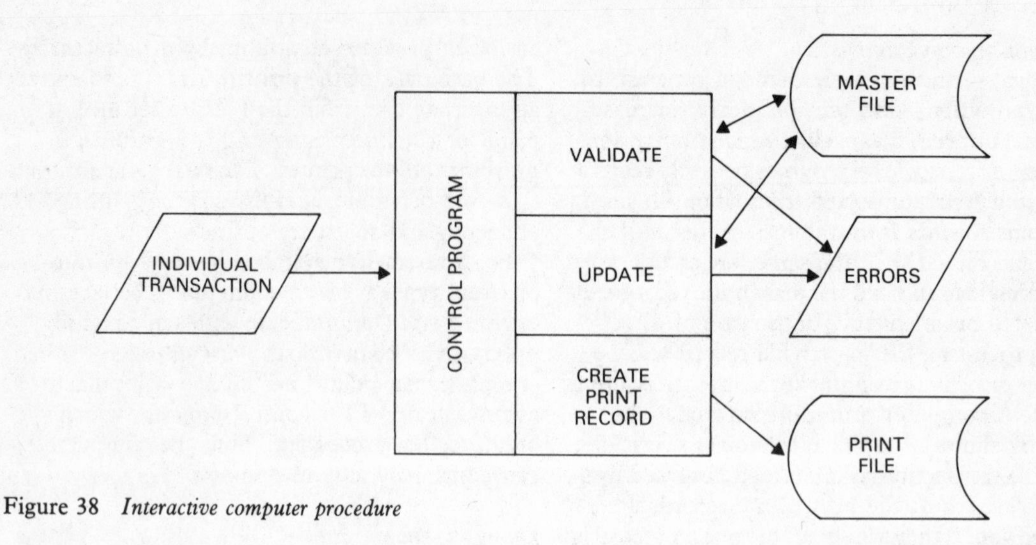

Figure 38 *Interactive computer procedure*

programming staff. However, one aspect of program design which is of particular concern to the analyst is the timing of procedures, because he has to be able to provide to the users and the computer operations manager fairly realistic estimates of the overall timing of the system. Large elements of this are concerned with external procedures in the user department – but the scheduling of jobs on the machine has to be based around accurate estimates of computer system time. The analyst therefore has to calculate, with the programming staff, set-up, processing and peripheral times for each of the programs involved in relation to the expected volume of data.

Finally, the systems analyst must ensure that adequate documentation passes between himself and the programmer, in terms of both the program (suite) specification initially and the program documents later. These will include some means of testing the acceptability of the programs before they are handed over, and evidence that the testing has been carried out.

User procedure design

User procedures are those activities carried out in the user department in the preparation of computer input (including in some cases data entry) and in the utilization of computer output (including response to interactive dialogues). They are clerical procedures but, because they interface with computer-based procedures, they tend to require formal definition. Their design is closely related to forms and dialogue design, computer procedure design and input and output design.

The major task for the systems analyst is to determine the flow of work to and from the computer and to document it for the benefit of the users. This involves the identification of tasks (including their boundaries) and their interrelationships and sequence, and the grouping of tasks together (ideally in a flexible way) into jobs which will provide interest, challenge and satisfaction to the incumbents. The flow may be documented using a clerical procedure flow chart as illustrated in Chapter 10. The aim of the clerical procedure flow chart is to provide a brief, clear, visual explanation to the users of the new procedures. But it also helps the analyst to check the efficiency, logicality, comprehensiveness and consistency of his procedure.

The work flow will be influenced by and will influence both the layout of the offices and the levels of resource provision. Decisions in these areas will probably be the outcome of work by organization and methods staff and personnel officers, but the analyst will offer advice. Minimum movement of work with maximum utilization of space in a pleasant environment are the aims of office layout design. Work measurement techniques will be used to determine staffing levels with the aim of having neither too many (to avoid boredom) nor too few (to avoid overwork) staff. The systems analyst may also be called upon to advise on organization structures suitable for the new procedures. The responsibilities of jobs and their reporting structure are inherent in the system design and the analyst has to place careful emphasis on appropriate communication paths and possibly on appropriate delegation strategies. In all of these issues, the departmental manager will have the executive responsibility but the nature of the system will be influential and so the analyst is bound to provide advice.

In situations which do not involve computer usage, user procedures gradually evolve with the users of the procedures being the people who make the changes. In computer-based systems, procedures are more formalized and tend to be designed from outside. A systems analyst has the task of deciding (or recommending) what others will do. The onus is on the systems analyst to promote user participation in design of the system so that their work procedures are what they want and not some external imposition. Job satisfaction for the user is just as important as (if not more important than) a beautifully efficient set of computer programs.

Forms and dialogue design

Forms and dialogues are linked together here because of the similarity in the principles behind their design. A form is a format on paper; a dialogue is a format on a screen. A format is used to guide a person in providing information in such a way that it can be correctly interpreted. The design of forms and dialogues is a task which requires great skill and cannot be treated casually; they need careful investigation and analysis before they can be designed and they need to be thoroughly tested afterwards. They also need to fit well into the system in which they are being used; they cannot be designed in isolation. Because both forms and dialogues are designed for computer users, there should be adequate user participation in the design. Too often the design may seem aesthetically pleasing to the systems analyst but be quite impractical for the users.

The main considerations in forms and dialogue design are the content, the layout, and the physical implementation.

Content

Content is concerned with two aspects of the format – the information which the format is intended to gather and the information which is necessary on the format to facilitate the gathering of this information accurately and efficiently. The systems analyst has to decide what information is needed; if the format is for an order, then the information may be date, order number, account number, product number, product description and quantity ordered; the items to be included will have been determined as part of the logical system design of inputs or outputs. The second stage is to decide what names wil be given to these pieces of information; they need to be unique and meaningful to the users so that they don't cause confusion. The third stage is to decide whether instructions are needed to help with the completion of the format, and whether these instructions should appear as part of the format or somewhere else (e.g. in a manual). The final stage is to decide whether the user of the format will have a free choice in filling in any particular section, or choice will be restricted to certain options.

Layout

Having decided the content, the designer has to attempt to optimize the layout of this content so as to make the format attractive and easy to use (without making it excessively large by going on to extra sheets of paper or extra screens). The layout is, of course, largely determined by the space available (the size of the form or screen) and the shape of that space. With VDUs there is usually no choice, but paper forms can have various sizes and shapes, though it is normal to use the standard paper sizes and to choose a horizontal or vertical layout.

The first stage in laying out the format is to estimate the size of the entries and to juggle with the space available for them until they fit in a reasonable way. There are some constraints on the juggling. The sequence of entries may be one constraint and this should always favour the person who will complete the document rather than the recipient. The size of the entry headings and instructions is often a problem; it is important not to allow these to determine the size of the entry space, and they should be adjusted to the size of the entry space. The use of multiple choice options can be space consuming because they need to be clearly separated to avoid confusion.

The second stage is to make use of underlining, lines and columns to highlight the different aspects of the format to which you wish to draw attention. Column lines help in identifying rows of figures; bold separating lines help to distinguish between separate pieces of data. On a paper form the use of different type sizes and faces can help in this; on a screen, the use of flashing areas and the cursor can assist in attracting the user's attention. The problem is eased with a screen format because the cursor will normally guide the operator through the format to ensure that the appropriate entries are made at the right location; checks will also, of course, be carried out on the entries because data entry is under program control.

Finally, the format needs to be rounded off with margins (especially on forms which are going to be photocopied or filed), a title and space for the reference number. A title is important as the identifier for the format which will be used by human beings. The reference number of the format (e.g. account number or order number or employee number) should be given a prominent position for checking and retrieval purposes.

Physical implementation

The physical implementation of the format is different for paper forms and screens. With paper formats, the designer has to choose the appropriate paper (bearing in mind paper size, weight, construction and colour), organize the printing (internally or externally) of the required number of copies, and, in some cases, select an appropriate make-up for the form (into pads or sets, into continuous stationery, with appropriate punchings and perforations, using carbons or no-carbon-required paper etc.). This can be very time consuming, and it is often handled by a specialist forms designer.

With screen formats, the analyst is more concerned with building flexibility into the design to allow use by different people. With a printed form, all users regardless of background, experience, intelligence etc. will be confronted by the same form; with a dialogue which is under computer control, some selection of the screen format appropriate to the operator can be made. Thus, a number of dialogues may be designed to achieve the same data collection purpose but for different people. The 'dedicated' operator needs less help than the 'casual' one; the 'active' operator will control the dialogue more than the 'passive' one who will tend to be instructed by the computer messages. Some users will need a lot of assistance, others will be intolerant of the system if they cannot immediately manipulate it.

Thus in dialogue design a more complex task is involved which aims to exploit the flexibility that the computer offers to the use of formats. The main two types of data collection dialogue are the form filling dialogue, where a format is set up on the screen and the user completes it as he would a paper form, and the menu selection dialogue, where the user is offered a choice of options and makes a selection from among the options to supply the required information.

Other factors which affect screen formats, which the analyst has to consider in their implementation, are the response time that is required of the computer by the operator (this can vary within a dialogue or between dialogues), the environment in which the VDU is being used, the types and intelligibility of error messages, and the relationship of the dialogue to the validation procedures which will form part of the input program.

The actual implementation of the dialogue will be carried out by the programming staff provided by the systems analyst with display layout forms and an indication, possibly in flow chart form, of the dialogue flow. Dialogue generators are special pieces of software which will assist the development of appropriate screen layouts etc.

A final comment that is very important is that forms and dialogues should be thoroughly tested in the field before they are used. What may seem acceptable to the analyst may prove ambiguous or confusing to the user. This should be avoided because the forms and/or dialogues are crucial tools in getting accurate input data to the computer and output data from it.

Security and controls

One of the essential features of any computer-based system is security – it must be protected against accidental or deliberate actions which lead to loss of confidentiality or loss of integrity. Most breaches of security are in the form of errors and omissions which are relatively small scale but can be costly in total; a few take the form of fraud and theft of information, and some involve the loss of computer facilities for a period.

The key to achieving security in computer-based systems lies not in procedural or technical arrangements, though these are very important and occupy most of the rest of this section, but in the overall organizational attitude to security. This means the attitude of the senior managers – because this will be reflected and imitated throughout the organization. If there is a lax attitude to security and confidentiality at the top, then this will pervade the whole organization.

In the past, information has been treated rather haphazardly, most of it being in the head or on bits of paper which would be impossible to correlate; computerization leads to formalization and this requires managers to be formal and explicit in their approach to information. In the past, security was achieved by disorganization and disintegration of information; as the computer makes information organized and integrated so security has to be examined and tested and considered in a way that was formerly unnecessary. It is management's job to carry out this rigorous analysis of security requirements.

From the analyst's viewpoint there are a variety of steps that should be taken in system design to ensure that systems are secure. The main ones are concerned with building controls into systems to try to prevent errors and omissions.

Controls

Control on computer input documents should be exercised to ensure that such documents (e.g. customer orders, time sheets, payments) are authentic; that they are accurately and completely filled in; and that none are lost during the receipt and checking procedures. When the documents have been converted into a computer readable form (i.e. punched cards, magnetic tape etc.), there must be checks (validation routines) to ensure that the conversion has been done correctly, and control totals should be printed.

Controls on output are equally important and the analyst will be concerned to define carefully:

1 The rules for printing reports (e.g. exception conditions);
2 The printing of confidential information;
3 Credibility tests (e.g. rejection of absurdly high payments, prevention of NIL invoices);
4 Authorization checks on documents to be sent to external parties;
5 Verification of master files by regular printout (e.g. comparison of stock balances with physical stock-takes, or computer prices with current price lists);
6 Distribution and storage of output reports.

The master files in any computer system contain

static data which is used frequently and so it is essential that this data is correct. Amendments to master files must therefore be carefully controlled; a record of all amendments required and changes to file made should be kept, and authorization for amendments should be at highest level. Equally, control must be kept over alterations to the file during normal processing. It is usual to produce batch or file control totals at every amendment or processing run. These control totals could be based on values, or number or records, or 'hash' totals or random fields which are treated as numeric though they may be alphabetic; sometimes all three are used. The important point here is checking the procedure that carries forward the control totals to the next run. The difference between the file controls from the previous run and those from the current run is compared with the control from the transaction or amendment file – and, if all is well, they should balance. Isolating of any errors is of course helped by maintaining small batches for control purposes.

Control over error correction is particularly worthy of emphasis for two reasons. First, errors need to be scrutinized carefully because they may well point to weaknesses in the security of the system. And second, the correction of errors is a particularly vulnerable part of a system and is frequently the location of fraudulent activity. Thus the analyst must design strict error correction procedures which involve full documentation of what has been done and which require authorization procedures.

Physical security

With regard to physical security, there are three main areas of concern. The first is the computer installation, i.e. the central hardware, peripherals and ancillary equipment, and functions associated with it. This has to be protected from environmental damage, malicious or accidental damage of hardware or files, and infiltration of systems. Of course many computer systems are now on-line, i.e. remote terminals are used to access central data, and the security problems in this situation are increased. The second area of concern is the provision of back-up in case the system does fail. And the third is the securing of insurance in case the precautionary measures taken are found to be inadequate. The security of the installation and acquisition of insurance are the responsibility of senior management. The systems analyst is concerned at a lower level with access control, file protection and system back-up.

Access control

The security threats in a terminal-based system include unauthorized use of terminals and use of unauthorized terminals to access files, and wire-tapping of data transmission lines. Techniques must therefore be devised to combat these threats. Four stages can be defined in controlling remote access to a central computer – authorization, identification, authentication and transmission – and the techniques can be examined in relation to each of these.

Authorization

One of the first stages in the design of a computer system should be agreement between all people involved on what data will be included in files and who will have access to it. Access to items, fields, records or files should then be authorized by senior management, and staff who are so authorized should be named. Care should be exercised in the authorization procedures to ensure that as few people as possible are entitled to access the files and that those who are entitled are reliable and responsible. This is the most important part of authorization – getting the administrative procedures correct.

Once people have access authorization, it is necessary to ensure that only they, and not anyone, are able to access the files. In a terminal-based on-line system the first level of control is access to the terminal itself, and the authorization procedures should make access available to those authorized and deny it to others. Terminals can be protected by being locked in rooms to which only authorized users have access; or they can be individually locked; or they can be built to include a badge reader which checks the credentials of the would-be user; or electronic combination locks can be fitted; or the user can be required to know certain codes or passwords in order to enter the system. If an unauthorized user attempts to use the terminal and is unable to overcome these protective barriers, the system should be able to lock him out. An authorized user as part of authorization procedures would be provided with the appropriate keys, badges, codes or passwords.

Identification and authentication

Having gained entry to the system the user must next identify himself. (This, of course, may have been done as part of the authorization procedure, e.g. by password or badge.) The problem here is achieving unique identification, and most of the current techniques can be circumvented by human ingenuity. Badges with punched or magnetic patterns can be stolen or duplicated; passwords can be heard, read or guessed. Work is under way at present to attempt to achieve unique identification via signatures, voice prints, finger prints and even facial appearance. These methods are at the research stage and it looks as though cost will keep them off-line for some time. So it is likely that current methods will remain for several years. The problem about current methods is the need to authenticate the identification – if uniqueness could be achieved the authentication problem would be resolved.

Authentication of the identity of the user is usually carried out by software techniques which are based on something the user knows, has, or is; in other words, the computer compares what it expects the user to know, have, or be, with what the user actually appears to know, have, or be, and decides to allow or deny access accordingly.

Transmission

The major weakness in terminal-based systems is in the tranmission link between the terminal and the central computer. The threat to security arises from the possibility of 'computer criminals' either eavesdropping on sensitive transmitted data by wire-tapping or picking up electromagnetic radiation or linking an unauthorized terminal into the terminal network. There are various steps that can be taken to detect such infiltration and to protect against it, such as shielding of lines over short distances, encryption (or enciphering) of transmitted data, and threat monitoring by software.

File protection

The basic hardware protection for files which is provided on most computers includes device interlocks, parity checking, file masking and write/permit rings. Device interlocks are a means of preventing interruption or termination of input or output once it has begun. This helps to avoid corruption of data by operator or hardware error.

Parity checking is a method of ensuring the validity of data as represented in the computer. Whenever data is read by an input device or from backing store a check is made to ensure that it hasn't been corrupted in storage or transmission.

File masking is a technique used with direct access devices to reject sets of commands. For example, all write instructions can be rejected to prevent corruption of an input file or all seek instructions to avoid unnecessary head movement. The integrity of the hardware protection can be protected by preventing subsequent attempts to set the file mask. On magnetic tape handlers, similar hardware protection is provided by write/permit rings which, if they are not attached to the device, will prevent it from being written to. This again is a method of protecting input files. Use of different types of hardware device in itself provides some security against unauthorized access. For example, in an on-line system the use of exchangeable discs for direct access storage means that all data is not always available. In fact certain files or even parts of files can be made available at certain times. Not only is this an economic approach to on-line storage, it also gives extra protection to files.

Hardware protection is provided in a multiprogramming environment to prevent users from disturbing the program or data of other users. They can also be prevented from untimely or improper output or input or halts by making certain instructions privileged, i.e. only to be executed by the operating system. Relocation registers, segmentation and paging are the techniques for protecting contiguous areas of computer memory from alteration by unauthorized programs. They do not, however, prevent such programs from accessing files on backing storage devices; this has to be handled by software.

Basic software protection of files consists of comprehensive checking of header and trailer labels on magnetic tapes and control areas on discs. File labels and control areas contain information such as file name, physical serial number, file generation number, date written, retention period, type of organization, maximum record length, number of records etc. From this, software can check that the correct version of the correct file is in use, that it is not being overwritten before its purge or expiry date, that no records exceed the correct length, that none have got lost, etc. These measures can safeguard the routine use of files.

Restart and recovery procedures are very important protection measures provided by software. With sequential files the problem is small because at any break in processing the file is in two distinct parts – part completely processed, and the remainder completely unprocessed. By noting the physical point at which the break occurred, the program can restart at that point. With randomly organized files the problem is much greater, because it is not possible to separate physically processed and unprocessed records. Recovery procedures must therefore include a log of either updating or updated records in the sequence in which they were processed. Processing can then be repeated by using this data; or can be restarted by identifying from the log the point at which the break occurred.

Software techniques can also be used to control levels of access to files. Access control can be imposed at several data levels down from complete file to individual field. *File level* is of course the easiest to implement; it only involves creating a decision table to define the conditions of use for each file, checking requests against the table, and allowing access to requests which meet the right criteria. There are two main problems about control at file level. First, it means that only one user can access a file at any given time, and so other users could be denied access for a considerable period. Second, it does not give a fine enough control over access to confidential data which is included in a file of mainly non-confidential data. The next level (in indexed files) would be an *index level*, whereby each entry in an index would have a decision table linked to it. This is really only workable where similar records are grouped together in the index or where each record has an entry in the index. *Record level* control would have the same effect as index level control in a sequential file for which no index is included. Here a decision table needs to be related to each record which would control access to the record. This means enormous overheads in terms of storing the decision tables and in processing time. *Group-item level* control is the next level, and this would define conditions of access to groups of related items of data, e.g. identification data in a personnel record. Because of the enormous overhead which would be involved in defining access conditions to group-items by record, it is probably necessary to apply this level of control to all records of the same type. This doesn't give very fine control, but it is probably adequate for most applications. The final level is *item level* control for specific items within a record, but the overhead for this type of access control would probably make it non-feasible for all but a very few applications. It should be noted, as one would expect, that the overhead in storing security decision tables and in processing time increases significantly as one moves from file towards item level control; but the security, in terms of allowing users access to smaller and smaller pieces of data, also increases significantly. Of course, all these levels of access control can be related to other security decisions. For example, should the file be available at all? Should it be only read, only written, or both read and written? Software can control access to confidential files (at some cost) in accordance with these types of decision – but it is perhaps worth noting that as the structure of a file becomes more complex, so the security overhead becomes greater. This especially applies, for example, to chained files.

Back-up

Despite all the security measures that are taken, no system can be absolutely secure, particularly against accidents, and so physical protection procedures must include a contingency plan to keep the system going in case of failure and to ensure that disruption of services is minimized. Such back-up needs to be provided in three areas, hardware, data, and system.

Hardware back-up
Hardware back-up involves ensuring that there is hardware available to continue running the system. Obviously there are different levels of need. In a magnetic tape installation, one tape deck can be kept free for use if one of the others fails; input to programs can be designed to use different media so that if the card reader breaks down, the paper tape reader can be used. These examples are simple methods of getting round minor hardware problems.

In a critical situation, where perhaps a fixed disk or the central processor becomes inoperable, a procedure must be designed before the failure occurs to cope with the situation. The procedure may be to do nothing and wait for repairs to be carried out. Or, in a more serious situation, it may be to negotiate with another organization who own a similar configuration to have the system run on their machine. If the installation is working 24 hours per day, or is doing mostly on-line work, then the back-up needs to be provided on-site and may well consist of duplication of critical equipment. Several installations now operate a dual processor system which will allow critical work to be switched from one central processor to the other when one fails or is closed down for maintenance.

Data back-up
In order to provide the duplicate files mentioned earlier, procedures have to be defined to ensure data back-up, i.e. to ensure that if a machine or program fails it is possible to rebuild files. All computer systems are based on cycles of work and on brought-forward and carried-forward files (in a batch system, the cycle is normally related to the batch; in an on-line system, the cycle is a period). A reconstruction method can therefore be based on retaining the input data and the brought-forward master file for a period of two or more cycles, so that the latest version of the master file can be reconstructed at any time. In a magnetic tape system one of the retained files can form the duplicate which is stored off-site; with direct-access devices, a magnetic tape copy has to be made at the end of the cycle. The duplicates, of course, need constant cyclical replacement.

With direct access devices, and especially on-line updating, the costs of providing data back-up have to be carefully considered. For example, if a disk file is copied before every tenth run of an updating program, the cost of producing the copy will be one-tenth of the cost of copying it every run, but the cost of reconstructing the file after a failure could be up to ten times as much. And with very large disk files, copying the whole file regularly may be impractical, and so procedures may have to be adopted whereby only portions of files are copied or only altered records are copied. In this case the copy master file will need to be periodically updated and this may provide a useful opportunity for file reorganization/restructuring. Many different methods can be used for providing data back-up and cost will usually be the determining factor, i.e. the value of back-up against the cost of copying and additional storage media. The important thing is that when program suites are designed, back-up and reconstruction procedures are included.

System back-up
Occasionally hardware and data back-up cannot provide the sort of cover that is required when something goes wrong with the computer system, and it may be necessary to continue operating without the files or a particular piece of hardware. It is necessary in this situation to have available (previously designed) procedures for operating the system in a different way until the fault has been put right. Sometimes this means devising emergency manual procedures; other times it means having a set of bypass procedures in the computer system (e.g. if a fixed disk went down, to be able to continue processing using tapes). In other words, the system designer needs to think about all aspects of system back-up and, depending on the importance of the system, emergency back-up procedures must be designed to meet all possible circumstances.

Overall guidelines
Finally, there are some overall guidelines which the systems analyst can employ to try to achieve security.

1 When systems involve the recording of sensitive data, the number and type of users of this data should be deliberately limited. Not only does this make security easier to achieve, it also aids in detecting breaches of security. Limitation of the number and type of terminals serves a similar purpose.

2 When hardware or software is being selected, particular weight should be placed on security features available. Security of systems will increase in importance as more data goes on to the computer; hardware and software which have poor security features will condemn the organization to poor security over several years.

3 A major safeguard in the design of systems to prevent either accidental failures or deliberate intrusion is the use of standard documents and methods. The benefits of standards are many, even though lots of organizations still seem unaware to them, but none more so than in the field of security. Control over the design and running of systems is easier because everyone involved can understand what each person is doing, programmers can less easily conceal fraudulent alterations to programs, and errors should be less frequent and easier to detect. There is a cost in standardization and it sometimes causes hostility from those required to follow the standards at first, but there is no question that the benefits outweigh the costs – if only from the viewpoint of security.

4 One of the main threats to security and reasons for privacy invasion is the centralization and integration of data. There is no doubt that advantages can come from this in terms of more contemporaneous, accurate, preprocessed, and flexible information, but at the same time the risks to security inherent in a massive collection of data are enormous. The system designer must weigh up these risks against the benefits, and must consider whether the fragmentation of files offers a more economic solution.

5 With regard to the storage, input and output of sensitive data, there are a number of simple rules that may be followed.

(a) Where possible all sensitive data should be encoded, with the codes only known by people authorized to see the data – for example, names should never be printed out.

(b) Whenever files are printed out, careful supervision should ensure that the number of copies is restricted and that they are distributed to appropriate personnel.

(c) Great care should be taken over printing out sensitive data in whatever form – when such programs are run, authorized personnel should be present, and after use the printouts should be shredded.

(d) All security measures, especially timings, personnel present, passwords etc. should be regularly changed.

(e) Great care should be exercised over the storage of duplicate copies of files and documents.

(f) If possible, it is preferable to keep statistical files (e.g. for planning purposes) rather than detailed, discrete, records about individuals containing historical information. Equally, emphasis should be laid on the storage of objective, verifiable facts rather than subjective opinions.

6 The systems designer bears the major responsibility for testing systems as a whole, and a great deal of thought has to be devoted to ensuring that all possible combinations of conditions are tested. Frequently the testing is restricted to the programs and this is a major cause of system failure. The designer must test the whole system – clerical procedures, data preparation and control procedures and operating procedures as well as computer programs – and he must test the security of the system as well as the accuracy of it.

7 Computer-based systems are extremely dependent on the people involved in them, and the successful system requires the commitment and confidence of these people. Much has been written recently on the need for the designer to encourage all the people involved to participate in the design process – and this need is particularly true with regard to security. Staff must understand the system and their role in it, and must be motivated to protect the system. A heavy responsibility in this area lies on the

shoulders of the system designer. He is the computer professional who provides the link between the technical safeguards and the administrative control procedures. He is the person who must advise and inspire managers, who must involve and assist auditors, who must educate those responsible for physical security, and who must build technical security measures into his computer system. This is a major responsibility which requires professional training, professional awareness and professional conduct.

Exercises

You work for a GCE examining board in an area which last year received examination entries in an average of ten papers each (equivalent to about five subjects) from 10,000 students. Between 1 and 10 August it is necessary to convert the marked papers into GCE results. Numbers are so great that the job cannot be done by manual means at an economic price (basically because it is a once-a-year job in the middle of the holiday season). So a computer system is required.

The system is to receive marked exam papers, to convert the marks into grades according to predefined rules, and to produce various reports, including:

(a) An analysis of marking by different examiners.
(b) An analysis of results by school and by subject with a comparison with the previous five years and with a listing for each school of all students whose grade in any subject is more than one grade higher or lower than the teacher's estimate.
(c) A letter to all students informing them of the results.
(d) A listing of payments due to each marker (calculated in relation to type of paper marked, grade of marker, and number of scripts marked).

It has been decided to design an outline system which can then be presented to computer manufacturers as a specification of outline requirements. You have therefore got to come up with some ideas as to how you see the system. These ideas should be presented under the following headings:

6.1 What input data is required (as detailed as possible)? Design a form to collect the data.
6.2 What is the most appropriate method of collecting this data and getting it into the computer? How can the data be validated?
6.3 What files will be needed by the system? What media will be best for storing the files?
6.4 What information will appear on the four output reports, and how should it be presented?
6.5 What processes are involved in the system? (Indicate the separate programs needed, including sorts etc.) What controls will be necessary?

To assist you in understanding this system, a data flow diagram of the logical requirements is shown in Figure 39. This is an extremely simplified model of what would happen in a real system.

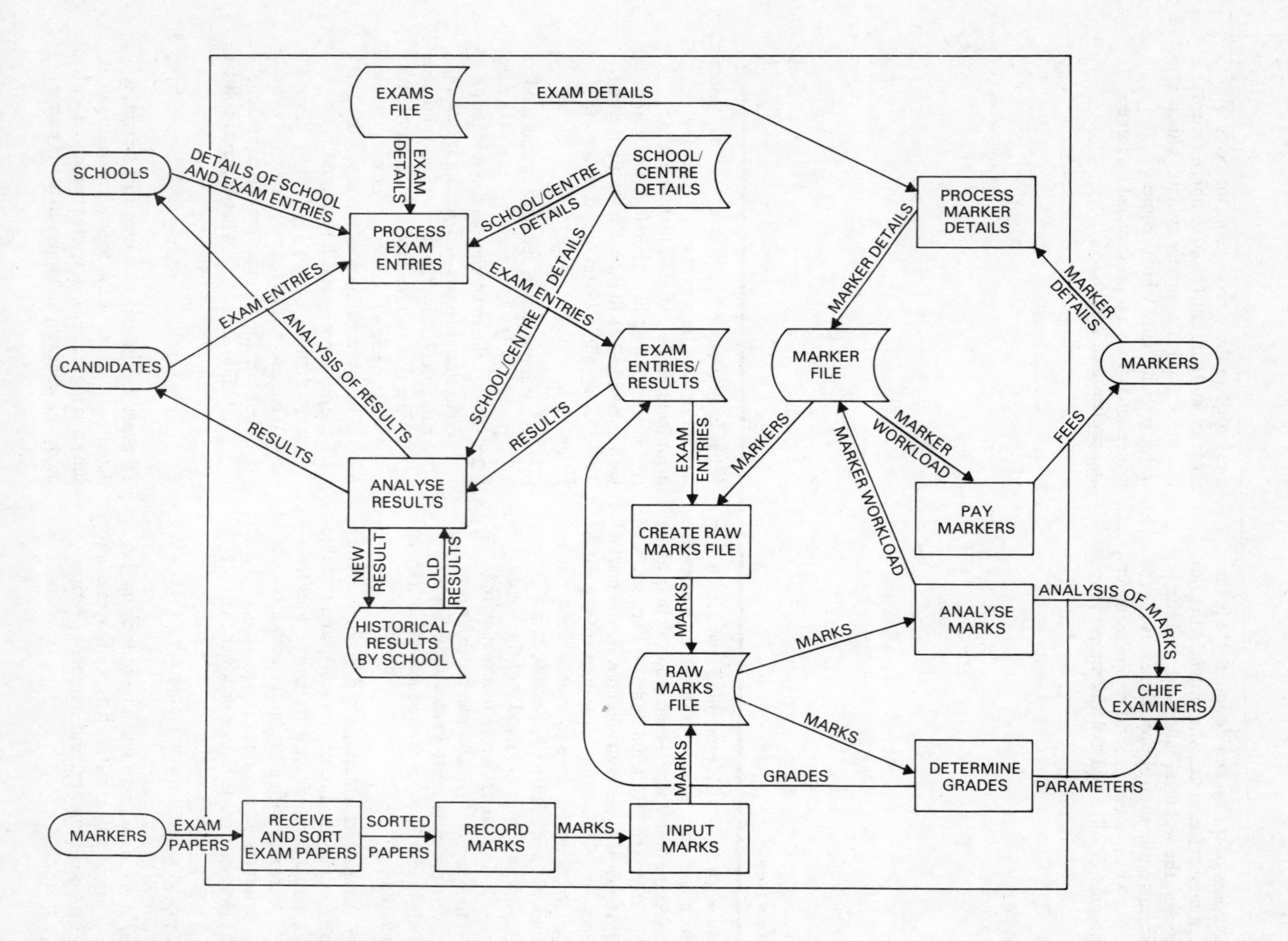

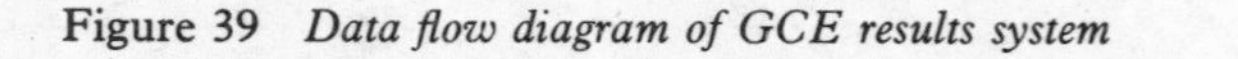
Figure 39 *Data flow diagram of GCE results system*

7 Implementation

Introduction

At a predetermined point in the detailed design the various specifications are 'frozen'; this means that they will remain unchanged until the new system had been implemented. The purpose of this it to allow the systems analyst and programmers to carry out development and documentation of programs and user procedures without fear of them being changed. Once the specificatons are frozen and the analyst is quite clear about the new system and its implications, the planning of implementation can begin – indeed, must begin, because there is a lot to plan. This planning activity will go in parallel with the writing and testing of programs and the writing of user manuals and computer operations manuals. Thus it will commence early in the project.

The implementation of a computer-based system is a large-scale activity. As will be seen from the time-scale chart in Figure 13, although the duration of the implementation is relatively short it involves large numbers of people, especially in the user departments. In fact, it can be said that the emphasis of the computerization project shifts away from the data processing department to the user departments, which face both unheaval in procedures and a greatly increased workload. Sometimes the implementation of a system is handled as a separate project in its own right, with a junior systems analyst responsible for guiding the users through the time of disturbance. Clearly, if the implementation is not properly planned, the disturbance can easily result in chaos.

Implementation planning

The planning of this stage should not be carried out by the systems analyst in isolation. Because of the nature of implementation, in which people have to be given instructions about their duties, it is essential that the planning is carried out by those with executive power, i.e. the line managers and supervisors of departments. The normal practice is to set up an implementation committee which is separate from the project team and which reports to the computer development steering committee. The implementation committee will consist of the managers of the departments affected, the systems analyst(s) involved, a representative of the personnel department and some representatives of user staff, and will be chaired by the most senior line manager. The systems analyst will offer advise to the committee and will assist in carrying out its decisions.

The committee will meet regularly during the planning of the implementation and very frequently (perhaps every two days) during the implementation itself, when it will be trying to resolve the various problems which arise from day to day. Initially its deliberations will be concerned with how to effect the implementation; later it will be concerned with sorting out specific crises.

The issues with which the committee has to deal include methods of implementation, staff selection and allocation, resources, and time-scale, but perhaps more important than any of these is the need for it to set up clear channels of communication and opportunities for consultation so that those affected have an opportunity to air their grievances.

Methods

Choice of methods of implementation is relevant to each of the activities shown on the chart in Figure 40. For example, there are different approaches to training, testing, file conversion and set-up and changeover, as will be described later. The committee has to decide which of the approaches is the most appropriate to the given circumstances.

Staff selection

Staff selection and allocation clearly is a critical aspect of implementation. Some staff will be relocated or redeployed; most staff will have

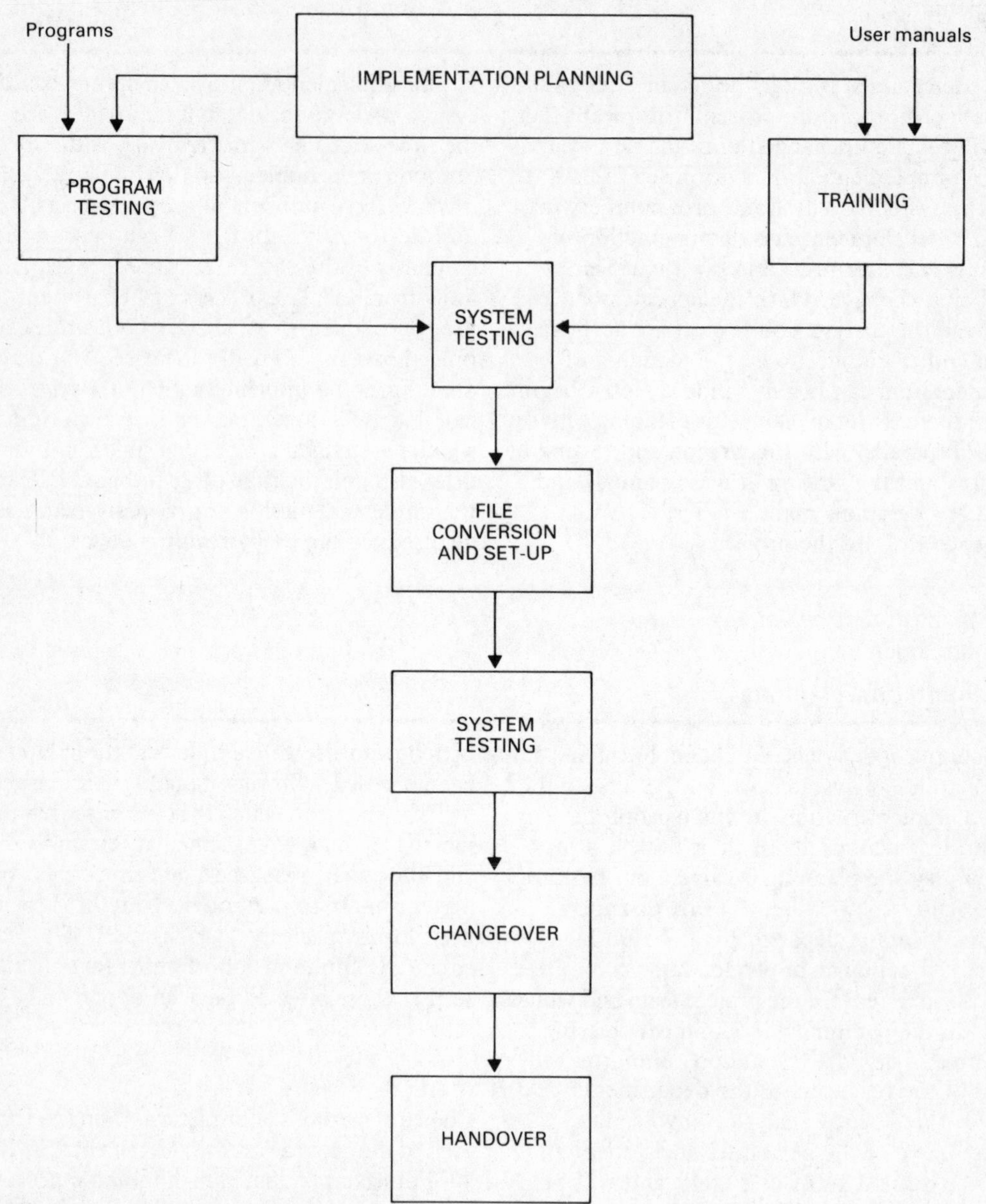

Figure 40 *Activities of implementation*

different jobs under the new system. It is no easy matter to decide who is going to do what but it has to be decided fairly early in the process – and certainly before training can begin. The line manager will normally make the choices (assisted by the systems analysts sometimes), advised by the personnel department (this is why the personnel department are represented on the implementation committee).

Resources

Resources are a crucial area of concern during the

implementation because the user staff at a time of great upheaval are often being asked to do extra duties. For example, training courses have to be attended, files have to be converted and checked, and sometimes systems (both old and new) need to be run in parallel. All of this has to be done whilst the user staff are keeping the old system in operation.

Time-scale

Finally, the time-scale has to be worked out. Time is invariably very precious during implementation. Files have to be not only converted but also put into use as soon as possible so that they don't become out of date. Testing has to be seen to be thorough but also swift so that other stages can be put into action. Above all, the changeover normally has to take place at a point in time (perhaps month end or year end) to gain the maximum benefits.

All of this points to the need for excellent planning and control to ensure the optimum use of limited time. And the planning must begin sufficiently early to allow full democratic discussion of the effect of the system on individuals.

Training

Training must, first of all, be distinguished from education. Training is about giving people skills; education is about giving them knowledge and helping them to adjust their attitudes. Education is a necessary complement to training but it should not take place at implementation time; this would be too late. Education should take place at the beginning of a project so that users are informed of what is likely to happen and, more importantly, so that they can contribute to the investigation and design stages. If the users have not been so educated and have not made such a contribution, the design is not likely to be very acceptable to them. Education sessions should be in the nature of computer appreciation courses (slanted in the direction of particular users) for staff at all levels; there should be lots of participation with opportunity for staff to argue about their doubts and grievances. By the time implementation arrives, the users should know more or less what the new system involves.

Training then is concerned with giving the users confidence in the new system by giving them the requisite new skills. Training is required for many different staff, including users (at all levels), computer operators, data preparation and data control staff.

Ideally training sessions should be conducted by managers and supervisors (aided by the systems analyst, and perhaps previously trained by the systems analyst) so that the message is passed on in the user's terms by people with whom there is some identity. (However good a rapport the systems analyst has managed to achieve, he is still an outsider.) The sessions should be short and regular rather than long and once-off in order to foster assimilation and learning. They should involve plenty of practice (at least 60 per cent) and demonstration (at least 20 per cent) and only a relatively small proportion of formal teaching or lecturing. Obviously the sessions must be supported by well-written and readily available user manuals. Job aids such as wall charts, notices, flow diagrams etc. should be made use of to provide visual assistance to the learning process.

In all of this, the role of the systems analyst is not so much to be the 'trainer' as to urge on management the need for full and adequate training for staff affected by the system. The users need this support and successful persuasion of management by the systems analyst would be a major contribution. The users' perception of the change will often be quite different from that of the systems analyst, and this is a good reason for encouraging user departments to do their own training.

System testing

The object of system testing is to ensure that the system is operating properly before the files are converted and set up on the computer in readiness for the changeover. At this stage the tests will normally be mainly carried out on the computer procedures (rather than the associated clerical procedures) using artificial data to attempt to check the accuracy of the programs. The testing of individual programs is the job of the programmer(s). When the programs are handed over to the systems analyst for system testing, they should be working exactly in accordance with the specifications provided by the analyst; in a sense, the program specification forms a contract between the analyst and programmer, and it is the latter's responsibility to produce a program which precisely meets the specification. Similarly the user system definition is a kind of contract between the user and the systems analyst, and the latter has the responsibility of ensuring that the system does what has been agreed.

It is good practice for the programs to be handed over to the analyst, with all the test data and results as part of the full program documentation. The analyst then has the opportunity to examine the results and to assess whether adequate program testing has been carried out. System test data should be completely separate and should become a permanent file which can be used to test the system after any subsequent changes to the system. System test data should test not only program logic but also volumes and working conditions. Response time, error handling, controls and rejection rates are, for example, of interest to the systems analyst (but not really to the programmer) with regard to their effect on overall system performance. The aim of the test should be to check all aspects of the system including the logic of the programs – although it is virtually impossible to check every possible pathway. The analyst will tend to concentrate on known error or exception conditions and examine the effect of these. Obviously he will be concerned that the standard program tests (e.g. oversize fields, incorrect formats, incorrect file, no matching records, file out of sequence etc.) are included in his test data, but his emphasis must be more on environmental factors. It should be possible in the system tests, for example, for the users to complete input documents or to enter data at the terminal, to access files, and to handle outputs so that they can get a feel for the speed of operation, the types of problems which can arise, and the nature of their interface with the system. This is why it is important for training of users to have taken place before system testing.

File conversion and set-up

File conversion is the activity of changing existing files into a form whereby they can be loaded on to the computer and are acceptable to the needs of the new system. File set-up is the process of loading and then checking the new files. File conversion needs to take place as late as possible in the development cycle – it would be pointless to convert files far in advance of the changeover date, because the greater the distance between conversion and changeover the more the changes that will have to be made to the converted files when they come into operation. Because of the need for a short interval between conversion and changeover, the conversion is a very critical activity in implementation and needs to be well planned and well executed.

It is almost a system project in its own right. The existing files have to be investigated and documented and a method has to be designed to do the conversion; this may involve forms design, clerical procedure design, training and even program specification and writing. Thus it can be an expensive activity. In addition, because the files have to be converted at the same time as the existing system is in operation, it is usually necessary to pay for part-time staff or for overtime for full-time staff to do the actual conversion. This will involve, normally, taking the existing file and transferring the relevant data from the records in it on to special forms which can then be punched for input to the computer. It is rarely possible for punching to be carried out directly from the existing records because of changes in codes, formats etc. in the new system. Thus, if one estimates the average size of customer file or product file at about 10,000 records, the size of the conversion task becomes clear; allowing half an hour for the transcription of each record on to the punching form, there are approximately 700 man-days of work involved and that doesn't include punching, computer time or checking.

In addition to the mere size of the exercise, file conversion is fraught with other difficulties which can be very awkward to overcome. For example, manual records tend to be not very accurate; this is because the users often hold the correct information in their heads and they do not bother to correct the written record. For the computer system, if inaccuracies are retained by the conversion process, errors will be magnified by the computer's inability to judge the accuracy of a particular piece of data.

Another problem is the need to edit existing records into the form required by the new system. If the customer code format is changed, then all the customer codes on all the records need to be changed. If the name and address are to be restricted to four lines of thirty characters, then someone has to decide on the format for each name and address. If the date of birth of each employee is to be recorded in a personnel system, it may be found that this was not the case in the old system and that several dates are missing which have to be collected. Lots of similar examples could be quoted.

A third problem is the location of the records which are to be converted. In certain situations (e.g. where a system involves the centralization of records which were formerly dispersed), the records may be located in several different places. The conversion of customer records for several different branches is a common example of this type of situation. Here, it may be necessary to bring the records together for the conversion process; if this is not done, then there is a danger that small differences will creep into the conversion process at each location. A slightly different problem (but with similar difficulties) occurs when the new records are to be a unitary combination of several previous records (e.g. for the system which crosses departmental boundaries). The new version of a customer file may include data from records previously held by the sales department (customer turnover details), the sales order office (customer credit-worthiness), the accounts department (state of customer's account) in addition to the basic information such as customer number, name and address.

The most difficult problem of all is the accessibility of live files. An existing file which is in constant use, such as a car-hire file, is required to be available for the purposes of the existing system. This means that two difficulties have to be tackled. First, time has to be found to do the conversion (probably out of normal working hours); and second, all the changes to the file which take place after the conversion and before the changeover need to be retained to be acted upon as soon as the new file is in operation. This can involve a lot of work and problems of control (i.e. ensuring that no transactions are lost or omitted).

Once the conversion has taken place (i.e. the data is in a form to be loaded on to the computer), the new file can be created. This is done using either the file amendment programs written for the new system (treating all the new records as insertions to the file) or specially written, one-off programs. Here the major problem is ensuring the accuracy of the conversion. Errors can occur at any of three points – at the transcription stage when data from existing records is written on to the file conversion forms, at the punching stage, or at the file creation stage. There is a clear need for careful validation of input, thorough control mechanisms to check that no records are lost, and checking (even if only on a sample basis) by user staff of a printed version of the computerized record. Ideally each record, once set up, should be inspected for accuracy.

The accuracy of the conversion cannot be overemphasized. If the system is to work correctly, obviously its files need to be set up and maintained correctly; 'clean' files are one of the prerequisites of effective data processing systems. Users will always believe that the files used in the manual system were perfect (even though it can usually be demonstrated quite easily that they were not); they must feel equally confident about the new files. If errors occur, invariably the users will say 'Well, they were always alright in our department.' The success of a system can often depend on the confidence which the users feel they can place on the files.

In the diagram in Figure 40, it was suggested that system testing would take place again after file conversion and set-up. To a certain extent, whether this happens or not tends to depend on the method of changeover chosen (and the extent to which this allows for system testing to continue). However, it can be said that system testing should be done again after file set-up, if only to test the files that have been created and to give users some experience of the actual data of the system. Not only do the procedures of a new system need to be tested (this has been done in the earlier system tests), but also the data (records, files) of the new system. The analyst should therefore try to ensure that thorough system testing takes place once the new files have been set up.

Changeover

Once the files have been converted (usually as soon as the files have been converted) the changeover from the old to the new system can commence, but this presupposes that various other preparations are also complete. For example, the users must have been trained in the changeover procedures as well as the new system procedures; all tests must have been completed to the satisfaction of users, management and computer staff; the various manuals about the system (including a changeover instruction manual) must be available to the relevant staff; all new equipment must have been commissioned and accepted; and the co-ordinating committee must be happy that the time-scale is operable. If all of this can be achieved by the target date and the new files have been cleared, then the changeover begins.

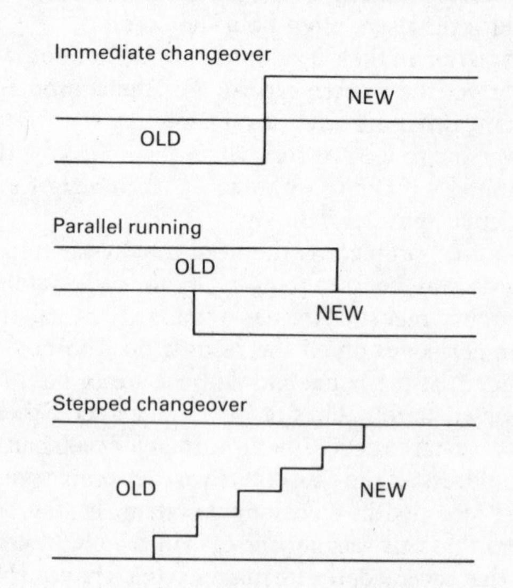

Figure 41 *Methods of changeover*

Immediate changeover

Changeover can be achieved in a variety of ways but the most common are 'immediate', 'parallel' or 'stepped' (see Figure 41). Immediate changeover is the name given to the method which leads to the complete replacement of the old system by the new at a point in time. The old system operates until the end of a week or a month, say, and at the beginning of the next week or month the new system begins. There is no fall-back position if the new system should be found to have unforeseen errors or problems after a few weeks or months, because the old system has not been in operation over that period. Thus, whilst immediate changeover is the simplest and least expensive method, it is also the most risky. It requires absolute confidence in the new system on the part of both users and computer staff and is better employed when the users have some previous experience of computerization than when they are complete novices. It tends to be used when the new system is not directly comparable with the old (and so no benefit is gained from continuing to operate the old), or when the time-scale of changeover is very tight, or when resources prevent a more thorough checking exercise such as is provided by parallel running. It is obviously essential with an immediate changeover that prior testing of the system has been exhaustive.

Parallel running

Parallel running is the direct alternative to an immediate changeover and involves a period during which the new system is run with the old, until everyone involved is happy that the new system is operating efficiently and effectively and the old system can be dropped. Obviously this is an expensive approach to changeover because more staff are needed to operate the systems in parallel and to investigate any discrepancies between the results of the two systems, but it does provide a fall-back position if the new system has problems. The advantages of parallel running lie in this delay in commitment and the extended opportunity for training staff and building up their confidence in the new system. It can however be difficult to persuade users that the old system is in error if differences in results between the systems arise. In any case, more and more computer-based systems are radically different from their clerical predecessors and so comparison between the two is often invalid.

Stepped changeover

The stepped changeover consists of a series of separate immediate changeovers on a small scale covering part of the system rather than the whole of it. The division into parts will normally be made on the basis of location (e.g. one depot changes over one month, another the next month, and so on) or subsystem (e.g. in a sales order processing system, order entry may be changed one month followed by invoicing the next month) or subfile (e.g. in a stock control system, the first third of the stock file may be changed over one week and the next third the following week). This method obviously affects the approach taken to file conversion. The benefits of a stepped changeover are that the burden of the changeover is reduced and spread over a longer period, and, more importantly, there is an opportunity to learn from the previous changeover and its problems before the next one takes place. The difficulties, however, lie in the need to control one system which is working in two different modes at any given point of the changeover period. It is also true to say that this approach tends to lead to a more protracted time-scale for changeover because subsequent 'steps' can be postponed until the previous ones have all been resolved.

Pilot running

A method of changeover which is often described but is strictly an extended testing situation is pilot running. The concept here is that data from a previous cycle of the old system is taken and run on the new system and the go-ahead to change to the new system is dependent on the approval of results of the pilot run. In practice this is a large-scale system test being carried out as a preliminary to an immediate changeover.

Handover

When the changeover is complete and the new system is in full operation, there should be a formal point of handover when support for the new system transfers from the systems analyst who has developed it to a maintenance group (this will be described in the next chapter). At this point in practice the users acknowledge that the system fully meets their requirements, and the systems analyst is able to withdraw to start a new project. The time from changeover to handover should be as short as possible because an idle systems analyst is a wasted resource. The handover should be formal so that those involved clearly understand the implications and so that the users can take on responsibility for their system; it should also be planned in advance and not allowed to be delayed. The users, however, must be satisfied that all is well.

Exercises

7.1 Write a short report from the viewpoint of a user describing what you would expect to happen during the implementation of a system and warning your manager of the potential problem areas.

7.2 Outline the difficulties which might occur in converting the hire purchase account records, currently stored in paper files at eighty branches of an electrical retailer, into a centralized computer-based file.

7.3 In the system mentioned in Exercise 7.2 it was decided to change over to the new system by parallel running. Discuss whether you think this was a wise decision and suggest some changeover problems which may have occurred.

8 System maintenance and review

System evolution

As soon as a system is in full operation (i.e. the changeover from the old to the new has been accomplished), it becomes subject to requirement for change. Computer-based systems, as indicated in the first chapter, are dynamic, open systems which have to adapt to changes in their environment. The changes can arise from a deliberate investigation of the performance of a system as measured against its objectives or from the discovery of errors or faults in the normal running of the system. The formal investigation of performance tends to be called a system review; and the making of changes to the system as a result of the recommendations of a review or as part of the everyday running of the systems is known as system maintenance. Maintenance of a system is part of the natural evolution of a dynamic, open system.

Sources of change

The basic reason why computer-based systems need to be changed is that they are concerned with processing data about the real world which is constantly changing. Businesses, for example, grow and expand or decline and contract; as the workload increases or declines it directly affects the volume of transactions to be handled, the timing of events, the resource requirements, the quality of work (and so on) in the data processing systems. Governments enact laws or dictate regulations; for example, sales order processing and invoicing systems had to be radically changed when the UK government changed from purchase tax to value added tax, and when decimal currency was introduced. Organizational policies change; a new product line may be introduced, requiring an entirely different costing system, or a payment system may be altered. People change; different managers have different ideas about the information they require for decision-making and as a result data processing systems have to be amended. Technology advances; as devices become more viable and less expensive, and as economic factors change, the technology on which a system is based can become obsolete; the growth of distributed processing, for example, has been facilitated by technological advances in telecommunications and microprocessors. All of these factors (and others which are peculiar to particular organizations) lead to changes in the data processing systems of organizations.

The required changes will tend to be identified in one of two ways – either as part of the normal operation of the system or as a result of specific formal review. In the former case, the people involved in operating the system may experience, over time, problems with or complaints about the system. It may be that timing is beginning to slip, or that output report formats are no longer appropriate, or that the response time is inadequate. On the other hand, there may be no problems but a recognition of a need for change because of direct external influences (such as government legislation). In either case, those involved in operating the system will initiate the various changes. With the system review, the changes are initiated from a variety of sources.

System review

The system review (sometimes called system audit) is a formal study of a system which covers similar ground to the feasibility study but in retrospect rather than prospect. The system review has three main aims: to assess whether the benefits of the system which were identified at the feasibility stage have been achieved (and, if not, why not); to bring to light areas within a system which can be improved by system modification; and to provide information about system development and design which may be beneficial to future projects (especially, for example, on the accuracy of estimates). The system review as a by-product should offer concrete evidence to management and users that the computer-based system does offer direct benefits to the organization. This will assist in diluting the general hostility towards computerization.

The system review should ideally be carried out on a regular basis (similar to a routine annual check-up) to establish whether the system continues to meet the needs of the organization. (This is particularly necessary in a rapidly changing environment.) It may be carried out by a systems analyst or a user – occasionally it will be carried out by a external consultant, perhaps from a firm of computer auditors – but it should of course involve users as much as possible because they are the people who can best assess the extent to which a system meets their needs. Normally the system review will result in a formal report to management and will cover the following aspects of the system:

The objectives of the system. Do these remain the same? If they have changed, what is the impact on system requirements? Does the system meet the stated objectives?

System effectiveness. Does the system do what is required? Are the various output reports useful and used? Is there sufficient responsiveness to change? Are error rates reasonable and contained? Is the response rate/turnround of the system adequate? How have increases in volume been handled?

System efficiency. Is the system operating efficiently in terms of resource utilization? Have all amendments been expeditiously and correctly implemented? Is the documentation up to date and accurate? Is the equipment fully utilized? Is the service provided to users at the level expected?

System acceptability. Are the users happy with the operation of the system? Has the old system been completely superseded? Is the level of absenteeism/staff turnover reasonable? Do staff approve of the changes made? Is the system easy to use?

System technology. Has advantage been taken of technological advances? Would significant improvements in efficiency result from adoption of new equipment or methods? How does the system fit into the corporate computer development plan?

System security. Are the quality assurance standards of the installation fully observed? Are the monitoring and control procedures sufficiently tight? Are the validation procedures sufficiently thorough? Are the auditors happy with system security?

Costs and benefits. The major concern of the system review is to assess whether the cost and benefit estimates that were produced in the feasibility study were accurate and whether, therefore, the system remains economically viable. The review provides an excellent opportunity to try to evaluate the intangible benefits which were identified at the feasibility stage.

In the light of all these factors the system review will draw some conclusions about whether the system is satisfactory. Particular emphasis will be placed on the relationship between expected costs and actual (especially costs which have arisen unexpectedly) and between planned and unplanned benefits. The requirement is for a system which remains viable – and which, hopefully, is more beneficial than was anticipated. The report will also make specific recommendations about desirable changes. These will be debated by the computer development steering committee and may result in proposals either for a complete investigation into the feasibility of a new system or for minor modifications to the existing one. Such modifications will take their place in the queue of amendments to the system.

Implementing the changes

Normally the implementation of changes to a system (i.e. system maintenance) is carried out by a system maintenance team which is distinct from the system development staff. Often the maintenance team (which will consist largely of programmers) reports to the computer operations manager; this avoids the conflict of priorities which might occur if the same members of staff were involved in both developing new systems and amending existing ones. The responsibility for maintenance activity should be clearly established in advance of it being required – and certainly after the handover point (described in the last chapter) the responsibility should no longer be with the analyst who originally designed the system. Maintenance work is often heavy on resources and most data processing departments would expect to employ at least as many staff purely on maintenance as they do on development of systems.

The major consideration during the maintenance phase of systems is to ensure that amendments are made correctly and at the right time and this involves the establishment of a rigorous set of amendment procedures. The first requirement is for a formal document which identifies the need for and nature of any proposed amendment. Normally this form will be completed by a member of the data processing department at the request of a user. Clearly only certain users will be allowed to initiate amendments to a system and each amendment request should be authorized by a senior manager (either of the user department or of the data processing department). The purpose of the amendment and its detailed specification should be completed on the form so that it can be carefully scrutinized before authorization. Amendments may affect forms or procedures or inputs or outputs (indeed, any part of the system), and so the specification must be perfectly clear and unambiguous. It should be at the same level of detail as the original system specification (e.g. sample screen layout for new screen rather than a narrative description).

A very important aspect of amendment procedures is the allocation of priority to amendments. Normally a distinction will be made between routine and mandatory changes; the latter are changes which are required because something has gone wrong and processing cannot continue without the amendment. In this situation, a programmer may have to work on the problem immediately but steps should be taken to specify and authorize the amendment in the usual way; otherwise there is a danger that no record will be made of what has been done.

Routine amendments tend to have a longer time-scale and can be fitted into a long-term schedule. They still need some further definition into date priority – to ensure that, where date is critical, amendments are done on time. Another aspect of this sequencing is the handling of concurrent amendments (i.e. simultaneous changes to the same program). There is no objection to this so long as one person is responsible for implementing all the changes to the given program.

Once the amendment has been documented it will be handed over to the appropriate person to be carried out. Part of this procedure is the updating of all the documentation affected. This may include program specifications, user manuals, computer operations instructions – often in several copies. It is essential that documentation is kept up to date so that those who are using manuals are not misled about the system with which they are dealing. Particular care must be taken over dating the amendments and making clear when they replace the old version of the system.

The amendment procedure must include very careful testing arrangements not only to ensure that the amendment has been done correctly but also to check that other parts of the system have not been corrupted during the amendment. This means that the original pack of test data, which was used to test the system when it first went into operation, should be available to test the amended system. This will help to ensure that the system retains its overall integrity. Often auditors will expect to see copies of amendments including the before – and after – copies of programs and files, and the test pack should be run to show the correctness of the changed program.

When all involved are happy that the amendment has been correctly completed, the new version of the program or file or form etc. can begin to be used. Good security procedures cannot be overemphasized. Using the right file or program at the right time is crucial to system integrity and steps have to be taken to control the versions in use. This equally applies at earlier stages, of course, when amendments are being made. The maintenance programmer must not be allowed to access live files or live programs which he could accidentally corrupt. He should make his amendments to a copy which can subsequently (once the amendment has been proved) be used as the live file or program.

A final aspect of maintenance is the cost of maintenance activity and where this should be allocated. If the users are to be educated about the cost of computer usage (and especially the cost of making what often seems to be a very trivial amendment), then they should be charged with the cost of changes. Not only will this cut out unnecessary amendments, but it will provide some incentive for the users to ensure that everything is done accurately and expeditiously. This philosophy to a certain extent depends on the overall system of charging used by the data processing department – if the users are not charged for the original development of a system it is slightly unrealistic to introduce charges at the maintenance stage. There are many arguments for and against charging for computer services.

One final comment. It will be seen from what has been said that maintenance is a costly part of the data processing function; it is therefore desirable that the work of maintenance should be made as easy to achieve as possible. This means that systems should be designed flexibly with maintenance in mind and the maintenance staff should have the opportunity to influence the design of systems to facilitate this.

Exercises

- **8.1** Compare and contrast the feasibility study (described in Chapter 4) and the system review.
- **8.2** Using the hints given in Chapter 6 on forms design, design an amendment specification and authorization form.
- **8.3** Suggest some arguments for and against charging for computer services within an organization.

9 Communication

Introduction

It should be clear from all that has been said so far that the systems analyst needs to be a good communicator. He has to deal with people at all levels in the organization, both those who are knowledgeable about computer technology and those who know nothing. He has to pass on ideas to other analysts in the project team, to programmers, to clerical and shop-floor staff, to senior managers, to people outside the organization (e.g. auditors, government officials, trade association officials), and to senior management. These ideas have to be presented persuasively, so that people are happy to accept them, and clearly, so that everyone knows clearly what is likely to happen and how it will affect them. Thus he is not concerned simply with passing on information but with presenting it in a way which will enable it to be understood and to be well received by those at the other end of the communication channel.

The main communication events that the systems analyst is engaged in are interviews (which have been discussed in Chapter 5), reports, presentations and meetings. The reports tend to fall into two categories, reports to management and technical operational reports. In the latter, especially, great advantage will be found in using standard methods of documentation. This will be elaborated upon later. For the moment we will concentrate on formal, written reports to management. Figure 12 gives an idea of some of the reports involved.

Written reports

Most written reports are a response to a specific request and will present in a formal way the findings of an investigation or design activity (e.g. a report on feasibility, or a survey of possible solutions to a problem). Thus, a writer of a report of this kind will normally have a clear definition of what is required, and will have authorization for the report. If this does not apply, then the writer is advised before commencing the report to establish exactly what is expected and to ensure that senior management have given authority for the report to be written. A report which is critical of a particular department's activities can be a very political document – and so its status must be clear to all involved. The systems analyst needs the protection of management authorization.

Preparation

The preparation for report writing involves the systems analyst in clearly understanding the answers to the following questions:

1. What is the report to cover? There is no value in a document which misses the point or which is full of irrelevancies. The content should be exactly what is required, no more and no less.
2. What is the purpose of the report? All reports are intended to achieve some specific purpose (occasionally, several purposes). Some will be intended to inform, others to persuade, others to present possible approaches. The purpose of the report will affect both style and content.
3. Who are the recipients of the report? The people who are going to read the report will influence the way in which it is written. For example, details of procedures will be inappropriate to senior management just as arguments about relative returns on investment will be of no interest to the sales ledger clerk.
4. When will the report be handed over? The timing of a report can greatly affect its acceptability. Normally it will be produced as part of an overall plan for a project and so it should meet that schedule. Related to this is the period of validity of the findings which should be made quite explicit. For example, cost estimates may be dependent on purchase of equipment within a particular period.

Answers to these questions are required before writing can commence and it is essential that the writer is clear about his objectives. One other consideration, before commencing to write, is the accuracy of material which is to be included. Even small errors in a report can throw suspicion on its validity – every effort must therefore be made to check details and sources before they appear in black and white.

Format

The format of a report will often be governed by the rules laid down within the organization, but it will probably be similar to this:

Title page. The report should have a front page giving the title of the report, the date of publication, the name of the author (and, possibly, the distribution list).

Contents list. The next page should provide a page-numbered contents list.

Summary of conclusions and recommendations. This should be a brief (ideally, one page) summarization of the main conclusions and recommendations of the report.

Introduction. This is the first section of the report and should define what the report sets out to achieve; it will normally include the terms of reference for the project, the approach taken, any modifications to the terms of reference, and a brief, descriptive walk-through of the report.

Findings. This is the main body of the report and may well consist of several sections or chapters which describe and analyse what has been discovered or what is being proposed. (The details of the findings should be relegated to an appendix.)

Conclusions and recommendations. Here will be found the conclusions (arising from the evaluation of findings) and the recommendations for further action.

Appendices. The appendices should include all the detailed technical data in the form of tables, lists, flow charts etc. together with, where appropriate, a glossary, a guide to further reading and an index.

Writing the report

Writing the report is of course not just a formality, although lots of time will have been spent previously in carrying out the study on which the report is based. The report needs to present information in an appropriate way so as to achieve its purpose. It is worth bearing in mind that reports are permanent documents and have an authoritative aura purely by virtue of appearing in printed form. This means that they need to be well written to gain the respect of a continuing readership.

The main emphasis in writing must be on pitching the material at the right level for the reader. Usually a report is written by a knowledgeable person for a less knowledgeable person, which means that carefully chosen introductory material is needed and the level of writing must be such that the reader can understand what is written. Beyond that the requirements of report writing are the same as for any other form of writing. The narrative must be clearly expressed, the facts should not be distorted but presented in a correct and fair way, and the material should be concisely constructed with attention to specific cases rather than to generalities. What is looked for in a report is objectivity – there is no harm in writing to sell or promote a particular argument so long as the alternatives and disadvantages are fairly presented. A very biased argument will tend to lose respect and acceptance. One of the conventions of report writing which is considered to lend objectivity to reports is to write impersonally (i.e. 'it is believed that . . .' rather than 'I believe . . .' or 'we believe . . .').

Style in report writing is a very subjective matter, but the following comments are offered for guidance (or argument!). It is preferable to keep reports as short as possible (whilst at the same time comprehensive) so as to encourage both reading and understanding. Short sentences are better than long ones, concrete terms better than abstract, and transitive active verbs better than intransitive passive. Adjectives should be kept to a minimum, and linking phrases avoided except where essential. It is important also to try to avoid ambiguity; for example, quantitatively imprecise words like 'some', 'most', 'a lot of', should be replaced where possible by the actual quantity or percentage. Having said all of that, the most important thing when writing is to write in a way that is most appropriate to the reader.

Visual aids are very useful in reports to increase rapid assimilation of information. Often a carefully constructed table or graph can convey as much as several pages of narrative and will be understood more quickly. Tables for classifying data for reference purposes, graphs to show relationships, ratios, break-even points etc., and charts such as bar charts, pie charts, histograms etc. are the most common form of visual aids, but illustrations of equipment, diagrams of layouts, flow charts, sample documents etc. all have important uses. Even cartoons are useful for breaking up long narrative passages and succinctly conveying a point (but they need to be drawn and used with care).

Packaging the report

Although the content of the report and the way it is structured and written are most important in terms of the effect of the report, the way it is presented or packaged can make a significant difference. If the initial reaction of the reader is distress at the appearance of the document or if the reader finds it difficult to negotiate the report then, however good the ideas, a poor impression has been created which will be difficult to counterbalance.

The first thing to think of is the cover and binding for the report because these convey the initial impact; the cover must look neat and professional to suggest to the reader that the content is similar. The inside pages should look uncluttered and well laid out to be easy on the eye and easy to follow. Neat typing with numbering and indentation of subsections is what is required. The printing also needs to produce clear copies which can easily be read, and perhaps the most appropriate method is offset-litho printing. In all of this the most important consideration is the standard approach of the organization; if there are organizational standards for covers, binding, typing, printing, then obviously these must be followed.

Summary

Reports are a lasting reflection of the systems analyst's work. They are read by senior management and, in some cases, by junior clerks. Not only do they need to reveal good ideas, sound judgement and professional expertise, but also they must be well presented to encourage understanding and acceptance. In many ways the systems analyst's reputation is built or destroyed by the quality of his written reports.

Presentations

Frequently a written report will need to be presented orally to the interested parties. In the case of a feasibility study this might be to the steering committee; in the case of a user manual it might be in the form of training sessions for user staff. In addition the analyst will frequently be called upon to give educational presentations about computer usage or particular applications. Indeed, education of the organization about computerization, its potential and its pitfalls can be seen as one of the analyst's main tasks.

A formal, oral presentation is an event which must be carefully planned. No one should be misled into thinking that it is very easy to give a talk off the cuff which will be appropriate to the needs of and well received by the audience. Very few people have the gift of extempore speaking; most of us have to prepare carefully and follow well-established rules in making presentations.

Preparation

The most important thing to realize is that the attitude of the presenter is more important to the quality of the presentation than techniques. For example, the person who is interested in his audience, who has taken the trouble to find out something about them, and who aims his talk at them as individuals, will be more successful than the person who is disinterested. Again, the presenter who knows his material, who is confident about it, and who believes in it, will be better received than the person who is not. Perhaps the most important factor in attitude to presentations is the extent to which the presenter is concerned about his own image rather than the needs of the audience. It is only natural for people, especially inexperienced speakers, to be nervous about the fact that they are the centre of attention during a presentation and that the audience may well be making judgements about them, but excessive concern about his own image will be detrimental to the speaker's interest in his audience. If he pays more attention to his appearance or to 'image-building' long words or to complex arguments to display his intellectual prowess than to the level of knowledge of the audience, then the presentation will almost certainly be a failure.

As with report writing, the preparation of a presentation requires the systems analyst to answer for himself certain questions:

1 Who will be in the audience? What will be their level of knowledge? Where will their sympathies lie? What are their individual needs from the presentation?
2 What material is to be presented? How much time is available? How much time is to be devoted to questions or practical work?
3 What is the aim of the presentation? (If it is to inform rather than to persuade, for example, the style needs to be quite different.)
4 Where will the presentation take place? What visual aid facilities are available? What are the acoustics like? What sort of services are conveniently provided?
5 What methods will be most appropriate to the presentation? Should discussion be encouraged? Should a case study/exercise be used? Would a film be helpful? Should slides be produced? Will an overhead projector suffice?

The answers to these sorts of questions tend to be based on experience of previous presentations, which emphasizes the need for the presenter to evaluate each presentation after it has finished to assess, via feedback from the user, whether it has achieved its aim and how it can be improved.

The common approach to preparing a talk is, in the light of the answers to the questions just listed, to sit down and put together as many ideas, thoughts, facts, opinions, illustrations etc. as seem relevant to the talk; then to sift through this material preserving the relevant and useful, and discarding the irrelevant and useless; next to decide the main points which it is aimed to convey (keeping these points to a minimum) and to work these into a theme for the talk; and finally to put a clear structure into the talk so that people find it easy to follow. Sometimes it helps to write out the talk in full; certainly this is advisable on the first few occasions. But the talk should always be given from notes rather than a full script; reading from a script will make the talk lifeless, boring and difficult to follow. The notes should be few, and so the speaker must know what he is going to say in support of the notes – but the use of notes will encourage the speaker to be more natural and more at ease with the audience (allowing some feedback to come from them to enable him to modify his talk as he goes along).

Structure

The structure of an oral presentation should be based on the knowledge that people find it very difficult to concentrate on and assimilate the spoken word. This has three implications: first, the presentation should be short; second, there should be a clear structure, and a certain amount of repetition to reinforce the communication of the message; and third, use should be made of visual aids to emphasize the main points of the talk. Thus a recommended structure is:

Opening. The first few words/sentences should be used to gain the listener's interest and attention and so should have some impact.

Signposts. The next thing is to make clear to the audience what the talk is about by giving a brief summary of its structure; the listeners should not be allowed to gain any false expectations.

Main points. Now comes the main body of the talk; in this the main points should be made to stand out (perhaps by the use of visuals) so that, if nothing else is learned, at least those points will be.

Summing up. The various points that have been

made should be regularly summarized and at the end of the talk an overall summary is appropriate.
Closing. The closing remarks should make clear exactly what you expect of the audience – in particular, if you are trying to persuade them to choose a particular course of action, the choice that you want them to make should be highlighted.

A much quoted phrase which emphasizes these points is, 'Tell 'em what you are going to say; say it; tell 'em what you have said.'

Speaking
A number of hints can be given to improve the performance of a speaker but, of course, individual style is more important than general hints. Perhaps the most important rule is 'audience contact' – the listeners must feel you are speaking to them as individuals and not *en masse*; this means that you should frequently look at the people to whom you are talking and try to catch their eye so that they are made to feel personally involved. The speaker's voice obviously must be clear and audible and this can be practised – it should also be natural and not contrived. The choice of language should aim for clarity of expression rather than creating a good impression. Often the audience can be distracted by the individual's mannerisms (both physical and verbal); people who wander about while talking or use the same phrase repeatedly are typical 'distractors'. These mannerisms should be avoided. The speaker should always face the audience to ensure that they can hear what he is saying. Stories and anecdotes can be very helpful both to illustrate points which are being made and to give relief in a long explanation (often it is helpful to provide people with images to aid their understanding); but the stories must be relevant and not lead the hearer off in some false direction. Perhaps the final hint is to be careful about timing; if you say your talk will last 20 minutes, then it should not go on for 40. Not only is it discourteous to make people late, it will lose you sympathy.

Visual aids
As was emphasized earlier, visual aids are very important to a successful presentation, because human beings take in more information through the eyes than through the ears. Visual aids can highlight the points of a presentation – or they can enable the audience to keep in touch with the development of the theme.

The most common visual aids are the blackboard, the flip chart and the overhead projector (OHP). Each has its advantages and disadvantages. The blackboard is flexible in use, but not appropriate to prepare in advance; it is dusty and, worst of all, it requires the speaker to turn his back to the audience. The flip chart (large sheets of paper which can be 'flipped' over) is a cheap and portable visual aid but needs to be carefully drawn on a large scale and cannot be reused if written on during the presentation. The overhead projector is the best solution. The transparencies are easy to produce by hand or by machine, they can be built up as a talk progresses, they can be written on in water soluble ink and then washed clean, they do not require a darkened room, and the speaker can face the audience. In addition, portable overhead projectors can be purchased.

Other visual aids are flannel boards, which allow the building of piecemeal diagrams; magnetic boards, which are similar in principle to flannel boards but make use of magnets instead of lint and felt; and film and slide projectors. The last two can be very useful to introduce pictorial images into a talk but they tend to require a darkened room (or a special daylight screen). Films in particular are fraught with danger because they can rarely be specific to a speaker's theme and so tend to be distracting; they should be used with caution and perhaps only for introductory material. The best visual aid of all is three dimensional, i.e. the object itself or a model of it which people can touch and examine.

When using visual aids, some points should be borne in mind. First, they should not be used for their own sake – in other words, they should only be used where they will help the audience. Second, they should not cause distraction – by failing to work, or being too noisy, or displaying spelling errors etc. Third, rooms darkened for any length of time lead to drowsiness and should be avoided.

Fourth, the visual should concentrate on main points and not trivialities; the OHP transparency should not be cluttered with information. Fifth, the speaker should never write and talk at the same time; the audience will not know which to follow, and will probably follow neither. Finally, do use visual aids!

Summary

For the systems analyst, presentations are an essential tool for communicating ideas to all the people with whom he has to deal. Attention can be drawn to the points which matter in a management report; staff can be educated in computer implications or trained in new skills; new systems can be sold to user staff. In all cases, the quality of presentation is crucial to both understanding and acceptance, and the systems analyst will find it necessary to become skilled in this area of activity above all others.

Meetings

A large proportion of the systems analyst's contact with users will be through formal meetings, for example, project committees, design and implementation working parties, departmental representative committees etc. He needs, therefore, to be adept at using these meetings to achieve his aims, whether these be to expedite an investigation, or to finalize agreement on the layout of a report, or to ensure that everyone is able to have a say in the approach to implementation.

The ingredients of successful meetings are:

1 The meeting's purpose should be abundantly clear to all participants; sometimes this means that the purpose has to be debated until all members have related themselves to it.
2 The number of people present at a meeting should be as small as possible in order to facilitate progress (though without making people feel they are not represented).
3 All members of a working party should be encouraged to see their role as putting the overall good before their own sectional interest; agreement rather than conflict should be the aim and meetings should be conducted in a good spirit (this is far more important than any set of rules).
4 Members of a committee have an onus to prepare themselves well for each meeting so as not to waste time; those who service the committee (often this will be the systems analyst) should ensure that members have every opportunity to prepare.

Meetings are usually run on formal lines with a chairman who controls the meeting and a secretary who services it. The chairman is usually a line manager; the systems analyst, being in an advisory position, will attend as an ordinary member (or occasionally as secretary).

The role of the various participants can be described as follows:

Chairman

The chairman is responsible for the overall planning of the meeting and its conduct. This involves discussion about who should be invited, when and where the meeting will take place, what matters will be placed on the agenda for discussion etc. These decisions will normally be made by the chairman in collaboration with the secretary. The chairman needs to study the people involved to forecast the way in which the meeting is likely to develop, and the subjects involved to keep the discussion on the right lines. During the meeting itself, the chairman's job is to control the progress of the meeting in terms of time-scales, participants' understanding of what is going on, and amount of participation. The chairman should encourage free discussion, inviting all members to contribute whilst restraining those who wish to contribute too much. At various points the chairman should provide feedback to the participants (and especially the secretary) about the state of the discussion, what has been agreed, and who is required to take action.

Secretary
The secretary is responsible for agreeing the agenda with the chairman and then circulating it and any supporting papers, once they have been printed, to the members of the committee or working party. The physical arrangements (accommodation, paper, travel arrangements etc.) need to be booked and checked before the meeting commences. During the meeting the secretary should take notes of the discussion and agreements and, from these notes, after the meeting produce a set of minutes of the meeting. These, after being agreed by the chairman, will then be circulated to the members as their record of the meeting. Included in the minutes will be an indication of action to be taken by various people; the secretary should try to ensure by following up once or twice before the next meeting that the various actions have been taken.

Members
The members of the committee or working party have an obligation to arrive at a meeting well prepared, i.e. having read and digested the papers and having planned their arguments/responses and proposals. During the meeting the aim should be a constructive and friendly discussion which may lead to disagreement but not bitterness. After a meeting, the members should carry out with all speed the actions which they have promised.

The systems analyst will find that these formal meetings are opportunities for educating users about the implications of computers (both their potential and their shortcomings); making sure that people understand what is going on is one of the systems analyst's responsibilities. He should be open and informative and should make helpful contributions to the meetings. Normally he will gain a lot from them in terms of understanding the people involved.

Standard documentation

A considerable part of the systems analyst's communication with users and other computer staff is of technical information (e.g. input, output, file and record specifications). In order to help people to understand this information, its presentation in a standard form on standard documents is necessary. If the users are confronted with different methods of presentation of the same information they will be confused.

What is meant by standard documentation?

The term 'standard documentation' is used to describe the forms which the systems analyst uses to collect information about an existing system and to design and communicate ideas about a proposed system. In essence this means that the forms provide agreed and accepted guidelines for carrying out particular documentation tasks. They obviously have to be related to standard methods and procedures of work. The objectives of using standard documents will be:

1 To allow communication of ideas and facts in a way which will overcome differences in interpretation and achieve speedier acceptance;
2 To provide a means of recording information in an economical format;
3 To offer a means of measuring performance and monitoring control on project development;
4 To reduce the chance of errors and omissions in collecting and presenting information (the forms can act as checklists for the inclusion of information).

Types of standard documentation

The standard forms that one would usually expect to see the systems analyst using fall into a number of categories. First, there are those forms associated with defining data items, records and files; these should enable the collection of information about data as it is used in an existing system in such a way that it can easily be referenced to the description of the new computer-based system. In other words, in the fact-finding stage, the analyst should be encouraged to gather information about size of data items, their content, frequency and interrelationships, about the number and growth of records in a file, about the frequency of change to records, etc.

Second, there are those forms associated with describing procedures. Usually narrative is quite inappropriate to the definition of procedures because it is too long winded and has a tendency to confuse. Instead flow charts, structured English and decision tables are normally used. (These are described in greater detail in the next chapter.) The important consideration here is that the charts should be designed in a standard way using standard symbols.

Third, there are those forms that should be used to record unstructured information such as the finding of an interview or the decisions of a meeting. The requirements here are, first, to ensure that the interview or meeting actually is recorded, and, second, to allow the participants to check that the record is acceptable to them. This means that no elaborate forms are required, but none the less a standard form is useful.

Encouraging the use of standard documentation

Systems analysts, like other people, are very reluctant form fillers, and so the incentives to use standard documents must be manifest. Normally this is achieved by the analysts' participation in the design and review of the documents, so that they can see and affect their usefulness and practicality. (In addition, of course, DP management will no doubt lay down a requirement for staff to use the standard documentation to aid project control.) Standard documents can assist in investigation, analysis and design tasks by providing a checklist of requirements, allowing multiple operation (by a project team) on the facts, and promoting quicker assimilation. They assist in the communication of information between people and between past, present and future. They help new trainees to learn what is required of them, to be productive earlier and to operate in a way that is compatible with their colleagues.

Standard documentation systems

Finally, it is worth commenting that while most organizations tend to develop their own standard documentation system, there are several schemes available nationally. Most of the computer manufacturers have recommended standards, and the National Computing Centre in the UK has a widely accepted manual of data processing documentation standards.

Exercises

A national chain of motor accessory shops is considering a number of methods of capturing data about items sold in the shops to provide input to a stock control system. The data processing manager has asked you to investigate three methods:

(a) keying of product codes into a cash register with cassette storage
(b) Kimball tags attached to the items, and
(c) bar-coded labels which can be read by a hand-held scanner.

9.1 Collect information about each of these methods.

9.2 Write a report to the DP manager presenting your findings and recommending the adoption of one of the methods.

9.3 Prepare notes for a talk to the shop managers explaining the method chosen, the reasons for the choice, and how the sytem will operate.

9.4 Produce a set of OHP visuals to illustrate the talk.

10 System documentation

Introduction

At different points in the book, various methods of documenting both computerized and non-computerized systems have been discussed. This chapter is intended to provide an overview of the methods together with a more detailed discussion of the specification of procedures.

In Chapter 5 the concept of a data flow diagram was introduced as the mechanism for providing a model of the user requirements of the system. It was said that the data flow diagram provides a link between analysis and design because the diagram can be used to depict both logical and physical aspects of a system. The data flow diagram is the key element of and provides a means of structuring the documentation of the system. Before we look again at the data flow diagram, a reminder is given of the symbols used for charting in this book (see Figure 42). The symbols are from the NCC *Data Processing Documentation Standards Manual*.

Some conventions of flow charting are:

1 Flow lines only cross in exceptional circumstances and, where they do, no logical interrelationship is implied.
2 Flow lines joining another flow line should always join at different points to avoid confusion.
3 The symbols should be of a standard size within a flow chart and be large enough to include text within the symbol; where there is not enough space for text, it should be recorded

Symbol (large)	*Symbol (small)*	*Meaning*
All of the first four symbols have a larger version so that more narrative can be entered if needed		Process
		Input/output or message
		File or data store
		Decision (in procedure flow chart)
		Movement (in procedure flow chart)
		Connector
		Terminal (in procedure flow chart) or external entity (in data flow diagram)

Figure 42 *Charting symbols*

on the facing page with some cross-reference in the symbol.

4 Symbols within a flow chart may be identified by a number to the top left of the symbol.

5 Symbols can be striped to provide extra information; a stripe at the top of a symbol will indicate that the symbol has a more detailed representation elsewhere and the references to this will be included in the symbol. In the following example, ORDPRINT refers to a procedure of that name.

ORDPRINT

6 A stripe in the bottom of an input/output or storage symbol can be used to indicate the medium, e.g. as an alternative to a specialized symbol:

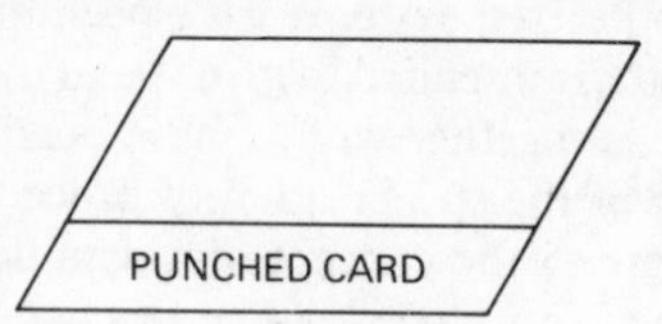

7 Connections of flow both within and across pages should be achieved by a connector symbol in which the appropriate symbol number should be written (if the symbols are numbered within pages, then the connector should show page number and symbol number).

Data flow diagrams

A data flow diagram is a tool for modelling or documenting the system as a whole in a structured way. It provides cross-references to more detailed documentation at lower levels in the hierarchy (some of which may be other data flow diagrams). The convention for a data flow diagram in this book is as in Figure 43.

The diagram which is Figure 43 can be expressed as:

> 'The system receives orders from customers, checks them against a product file (to see if the item exists and there are some in stock) and against a customer file (to check that the customer exists and is credit-worthy), updates the stock level on the product file, and sends the goods ordered with an invoice to the customer.'

This particular data flow diagram tells us very little about the working of the system but gives us an overall idea of its scope. For more detail it needs to be exploded or decomposed. The process box in this case could be decomposed as in Figure 44.

Each of the process boxes in Figure 44 could be further decomposed until the required level of detail is reached.

Each data flow diagram should be enclosed in a box to indicate which files, processes and messages are unique to that system or subsystem. Any file, for example, which is used also by another subsystem or process would appear outside the box. A dashed line on a data flow diagram can be used to draw a boundary around certain symbols, such as those which are to be computerized, those which form part of a priority development, or those which are to form the activities of a particular user department or section.

Once the data flow diagrams have been hierarchically decomposed to an adequate level of detail (which will vary from process to process), the various symbols can be cross-referenced to a more detailed specification (using where appropriate the top stripe of the symbol to enclose the cross-reference). Thus each message symbol will have written on it the name of the document which specifies the message; each file symbol will have within it the name of the document which specifies that file; and each process box will be cross-referenced to the more detailed specification of its logic. The names used in the symbols should be brief; the names in

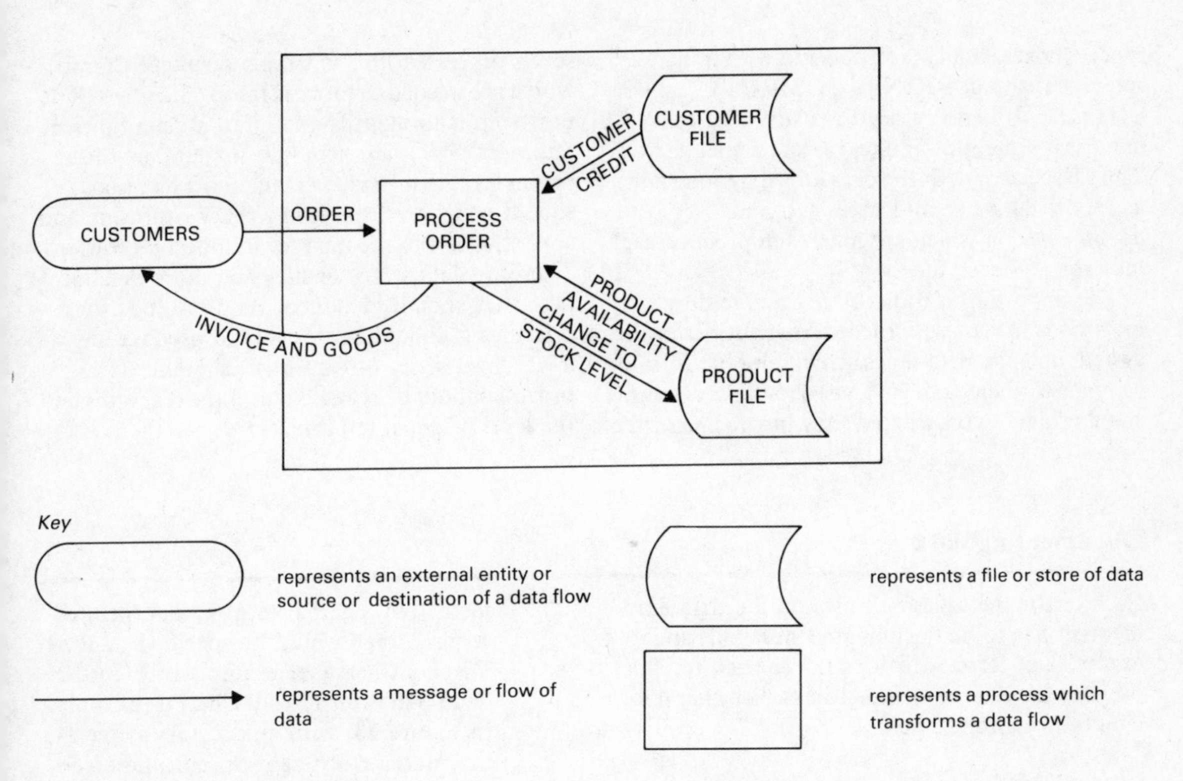

Figure 43 *Simple data flow diagram*

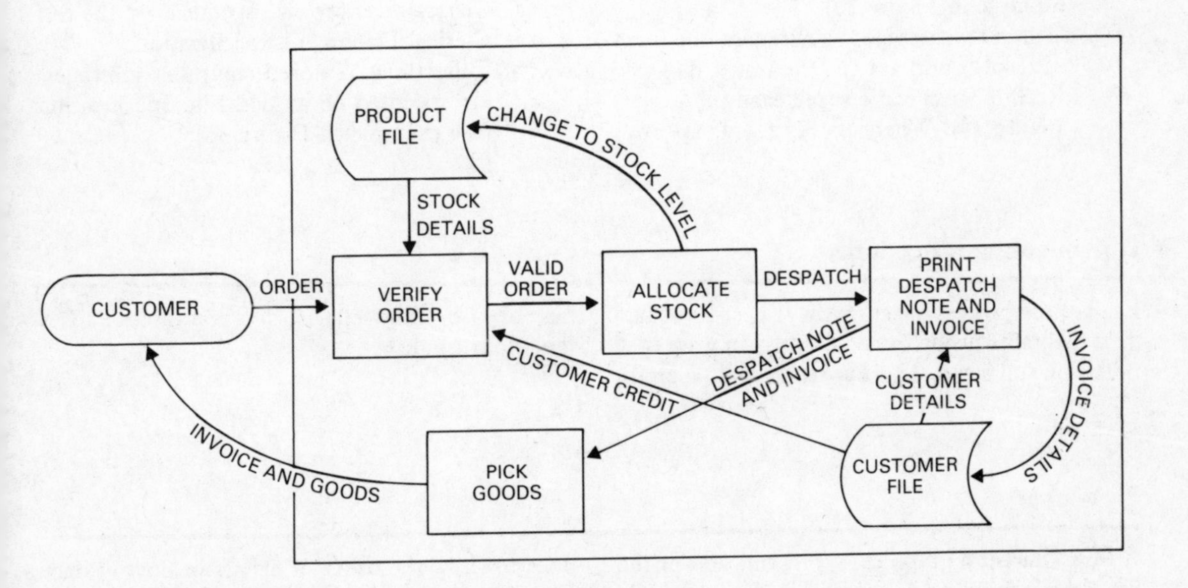

Figure 44 *Explosion of the data flow diagram in Figure 43*

process boxes should always begin with transitive, active verbs (e.g. PRINT, COMPUTE, CHECK). It is necessary to ensure that every name is unique; this applies to data names as well. Thus associated with the data flow diagram shown in Figure 44 we should expect to find documentation which specifies each process, each message and each file.

The approach to constructing a data flow diagram is to start with the external entities of the system and the data flows which emerge from or are sent to them; a high-level data flow diagram should then be constructed showing no more than seven processes (this is simply a rule of thumb) which are needed to process those data flows. The next step is to identify the internal data flows or messages which are necessary to support these processes; at this stage errors and exceptions should be ignored since they cause confusion and are better left to the next level of decomposition. Then the data stores or files and more detailed processes should be added. As a result of these activities it should be possible to produce a lower-level series of data flow diagrams. This process should be repeated until the diagrams and the level of detail are satisfactory (to the user?).

Documenting data

Each of the messages and files on the data flow diagram has to be documented in detail, and examples of forms which could be used for this purpose have been included in earlier chapters. The forms involved are:

1 Messages
- (a) Clerical messages are specified on a S41 Clerical Document Specification (see example in Figure 20).
- (b) Computer messages usually take the form of records and are specified on a S44 Record Specification (see example in Figure 34). Where these records are formatted together (e.g. screen, printed report), they would be specified on a S47 Display Chart (see example in Figure 29) or an S46 Print Layout Chart (see example in Figure 28) with an accompanying S43 Computer Document Specification (see example in Figure 30).

2 Files
- (a) Clerical files are also specified on the S41 Clerical Document Specification.
- (b) Collections of stored computer messages are specified on an S42 File Specification (see example in Figure 33).

Documenting procedures

Each of the process boxes on the lowest-level data flow diagrams needs to be specified in greater precision and detail. There are three main tools for this purpose – flow charts, decision tables and structured English.

Flow charts

A flow chart is a graphical representation of the procedures involved in a system; in it symbols are used to represent data and operations and flow lines to show the order of events or flow of data. Flow charts can be used to show a variety of different procedures at different levels. At the highest level within a system, we have the system flow chart which shows the overall activities of the system by department. Below the system flow chart and at a lower level of detail is on the one hand the clerical procedure flow chart, which depicts the user procedures, and on the other hand the computer run chart, which shows in outline the computer procedures including inputs, files and outputs. The computer run chart can be defined in more detail by the use of a computer procedure chart. All of these charts can be used in the investigation of a system (to record findings), in the design of a system (to record a proposal), and in the specification of a new system (to record the operational procedures). They are used not only as a recording tool but also as an analytical tool to allow the systems analyst to check the consistency and correctness of what happens within a procedure.

Let us look at some flow charts.

The system which is to be flow charted can be described in narrative form as follows:

> 'Customer orders arrive, by post and by telephone, in the XYZ Ltd sales office. Posted orders are sent straight on to the order transcription office, while telephone orders are sent in batches periodically. In the order transcription office, each order is transcribed on to a computer order form. The customer account number, looked up from a card file, and product codes, obtained from a code list, are added. If there are more than ten items ordered, then another computer order form has to be used, and so on. The original orders are filed in the customer order file. The computer order forms are batched together at 11.30 a.m. and 3.30 p.m. and sent to the machine section for calculation of control totals and completion of a batch header form. This is attached to the batch of computer order forms and sent to data control section where the batch number and control totals are logged, before the batch is sent (with the batch header form) to the punch room. A card is punched for each item on each order and for the header form. The card deck is then passed with the batch of documents to the data control section where it is logged again, checked for completeness and then passed to the computer room. The documents are retained by data control.
>
> The cards are fed in batches into the computer and validated. The validation program prints out any errors in the cards or the batch controls and rejects a whole of batch of cards if an error is found. Valid batches are written on to magnetic tape and continue to be processed. Meanwhile the error printout and batch of rejected cards are returned to the data control section. There the errors are examined and corrected if possible; if they cannot be corrected the computer order form in question is sent back to the order transcription office for correction. Once all the corrections have been made, they are written on the error printout and passed back to the punch room for punching. They are returned to data control and checked before the batch is submitted for running at the next validation run. The computer produces for valid orders a five-part warehouse despatch set, on continuous stationery, which is passed to data control section for guillotining. The guillotined despatch sets are then sorted into urgent, priority and ordinary despatches and sent in three separate piles to the warehouse manager.'

A data flow diagram of this system might appear as in Figure 45. Each of the processes shown on this data flow diagram is striped. This indicates that the process is defined in greater detail by another specification (which might be a clerical procedure flow chart or a decision table or structured English). So in this case in our specification of the complete system we should find additional procedure specifications for SOPR, OTOPR1, MSPR, DCPR3, DATAPREP, DCPR4, and COMP. It might well be that the next level of specification will be another data flow diagram. So DATAPREP may be a data flow diagram which has on it three processes DCPR1, PRPR, and DCPR2. Let us now look at a system flow chart for that same procedure (it is shown in Figure 46).

The system flow chart is an alternative representation of the data flow diagram showing the physical implementation of the system by

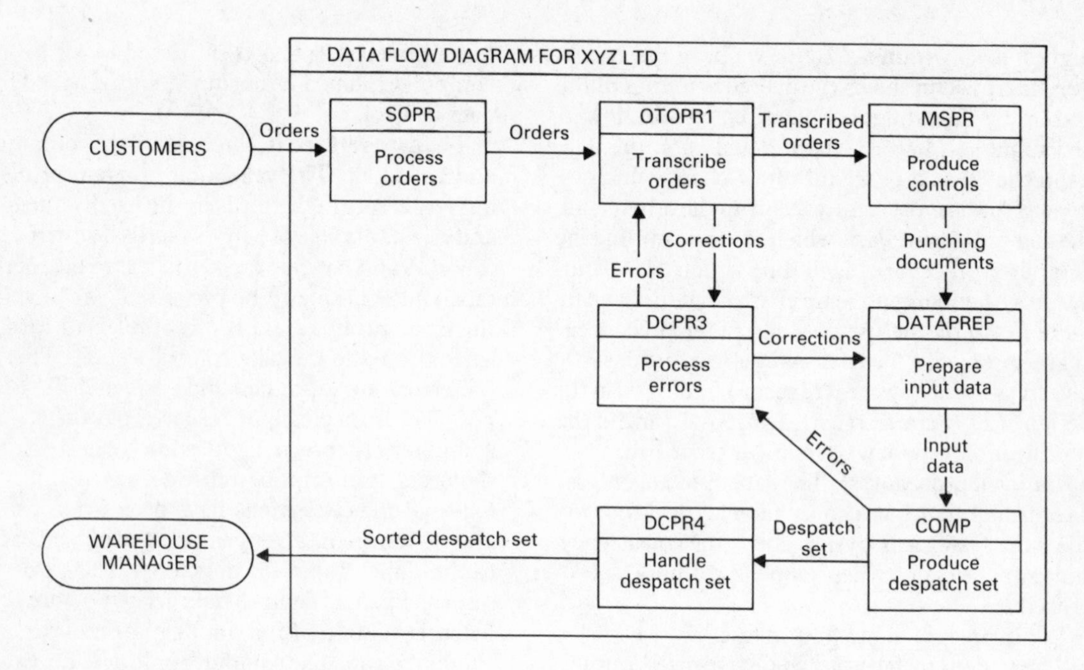

Figure 45 *Data flow diagram for XYZ Ltd*

departments in terms of the flow of activities. It uses the same cross-references to more detailed procedure specifications as the data flow diagram but shows the processes as a sequence of activities, from the arrival of customer orders in the sales office. Each new document which is created is shown, and each overall process in a department/section is represented by one striped process box. The reference in the stripe is to the more detailed flow chart which describes that process. Thus SOPR, OTOPR1 and 2, MSPR, DCPR1, 2, 3 and 4, and PRPR refer to clerical procedure flow charts which define those procedures; and COMP refers to the computer run chart for the program suite which produces the two outputs. Thus the system flow chart is intended to offer an overview of the system and the department/section involved. It shows the flow through the sales office to the order transcription office, to the machine section, to data control and the punch room and eventually to the computer.

Having established the overall procedure we can next turn to the details, and Figure 47 shows examples of clerical procedure flow charts for two of the processes shown on the system flow chart and data flow diagram – MSPR and DCPR4. The clerical procedure flow chart shows the documents involved in more detail (e.g. the despatch set is shown in its five parts) and all of the processes involved in relation to the documents. In the MSPR example it shows that the computer order forms originate from procedure OTOPR1, which would be separately defined, and are used to produce control totals which are written on the batch header form; the documents are then grouped together and sent to procedure DCPR1. The whole of the clerical procedure could be drawn on one flow chart but the author believes that the principle of showing the system flow chart by department and then producing clerical procedure flow charts for the activities of each department offers a clear structure. Clerical procedure flow charts can be drawn to place emphasis on the documents, the procedures or the location. Emphasis here is on the processes carried out on each document; thus it should be possible to trace a document from its original to its ultimate destination through the flow charts.

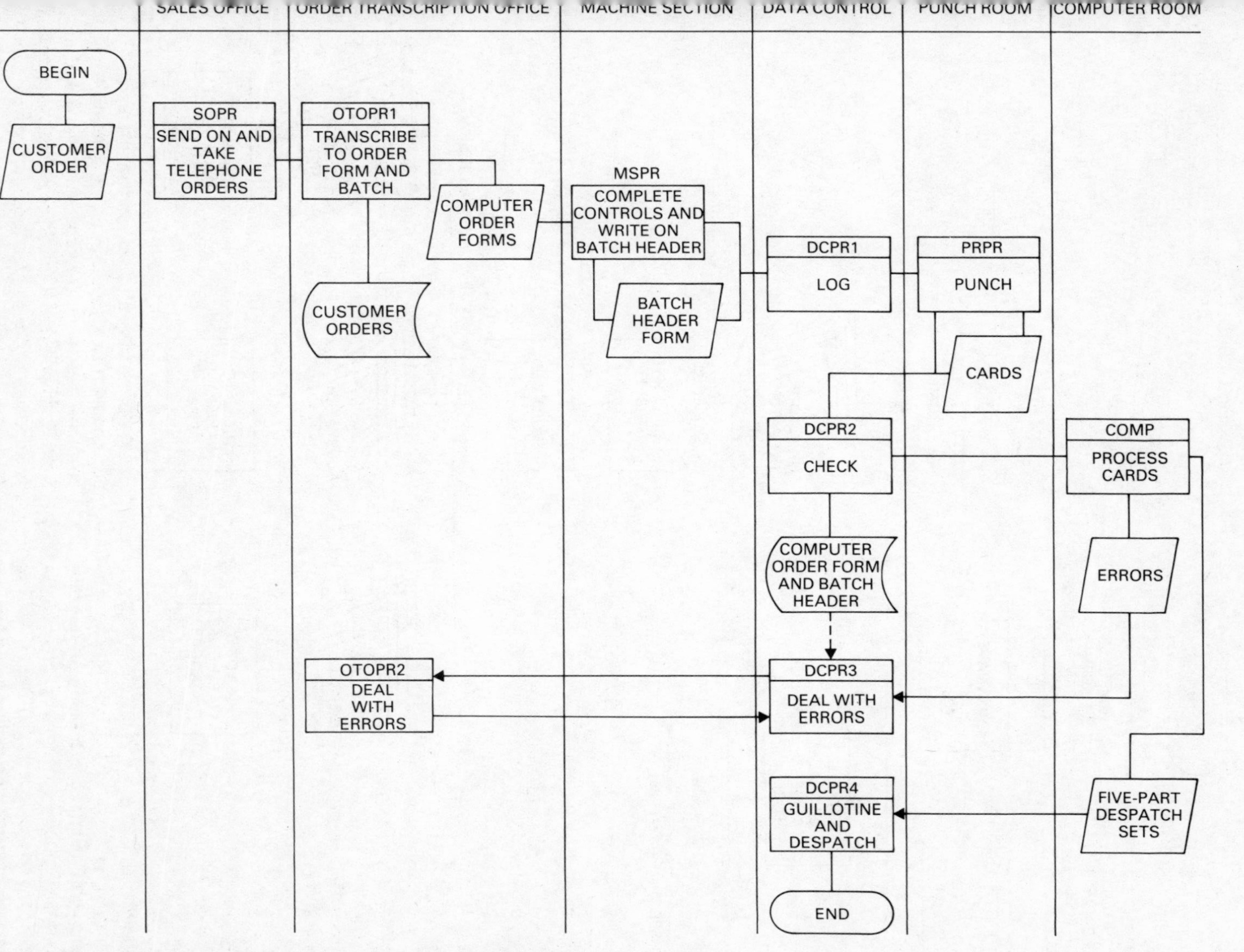

Figure 46 *Overall system flow chart for XYZ Ltd – order processing*

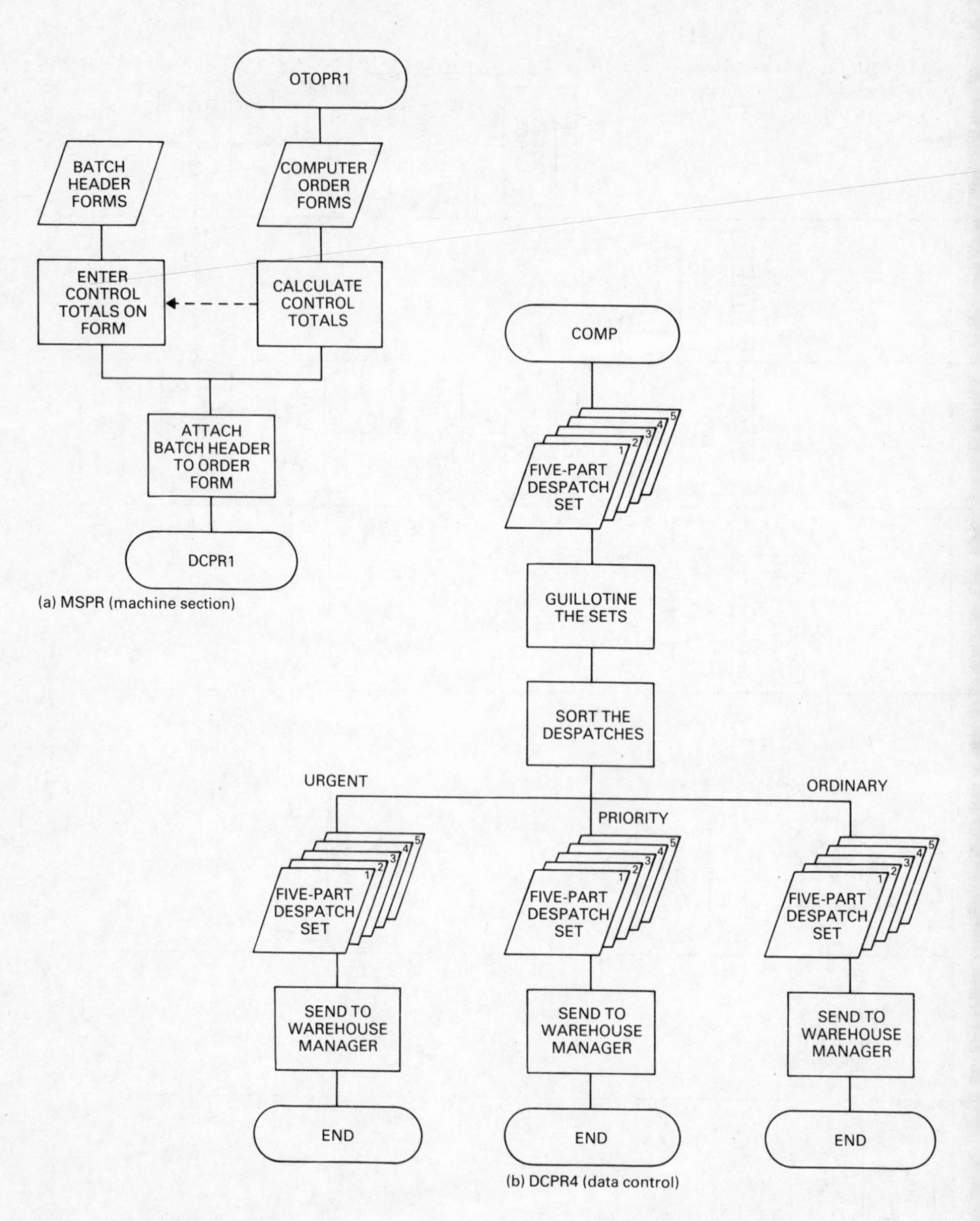

Figure 47 *Examples of clerical procedure flow charts*

On both the data flow diagram and the system flow chart the computer procedures are represented by the box COMP, and this is expanded into a computer run chart in Figure 48.

This shows the cards being fed into the system, and the outputs from the system which are used in the clerical procedures, i.e. the report on errors and controls which are used in DCPR3 and the

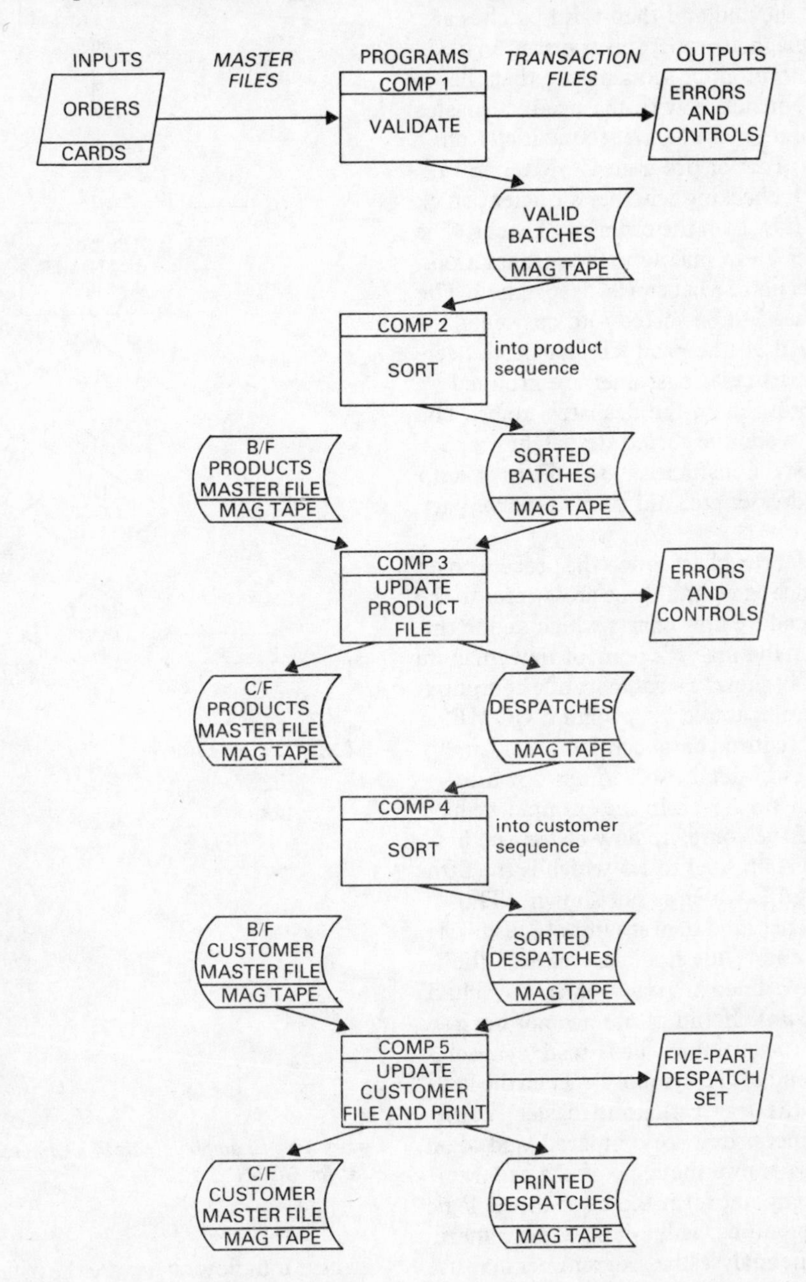

Figure 48 *Computer run chart for XYZ Ltd – order processing*

despatch sets which are worked on in DCPR4 (shown in more detail in the previous clerical procedure flow chart). In addition the run chart shows how the outputs are produced. The cards are read and validated and then valid batches are written on to magnetic tape; this transaction file is then sorted into product sequence so that the records can be matched with the product master records which are stored serial-sequentially on magnetic tape. Part of program COMP3 will be concerned with checking whether sufficient stock is available; if it is, then the order can be met. The program writes on to magnetic tape transactions which show details of what can be despatched. The records next need to be sorted into customer sequence so that all the products that have been ordered by a particular customer are grouped together for printing on the despatch notes. The final program reads the sorted despatches tape, matches against the customer master file to pick up names and addresses etc. and prints the five-part despatch sets.

The computer run chart shows the process boxes with stripes once again and these cross-refer to the computer procedure flow charts which define the programs, from the analyst's point of view, in more detail. Figure 49 shows a small part of a computer procedure chart example for program COMP5. Computer procedure charts can be drawn in varying degrees of detail, sometimes down to program instruction level. In the example we have only page 2 of the complete flow chart; the first process box is connected to 1.5 which is the fifth symbol on page 1 (which is not shown). The despatch file is read and then at symbol 2 tested for end of file; if end of file has been reached the procedures are defined on page 8 of the flow chart (again not shown). If end of file has not been reached, then the customer file is read (symbol 3) and tested for end of file (symbol 4). Then the keys of the two records (transaction and master), in this case the customer codes, are compared; and so on. This flow chart shows the logic of the program. This may or may not form the basis of the logic which the programmer will follow. It is primarily for the use of the analyst; the programmer may use some form of structure diagram to design the program.

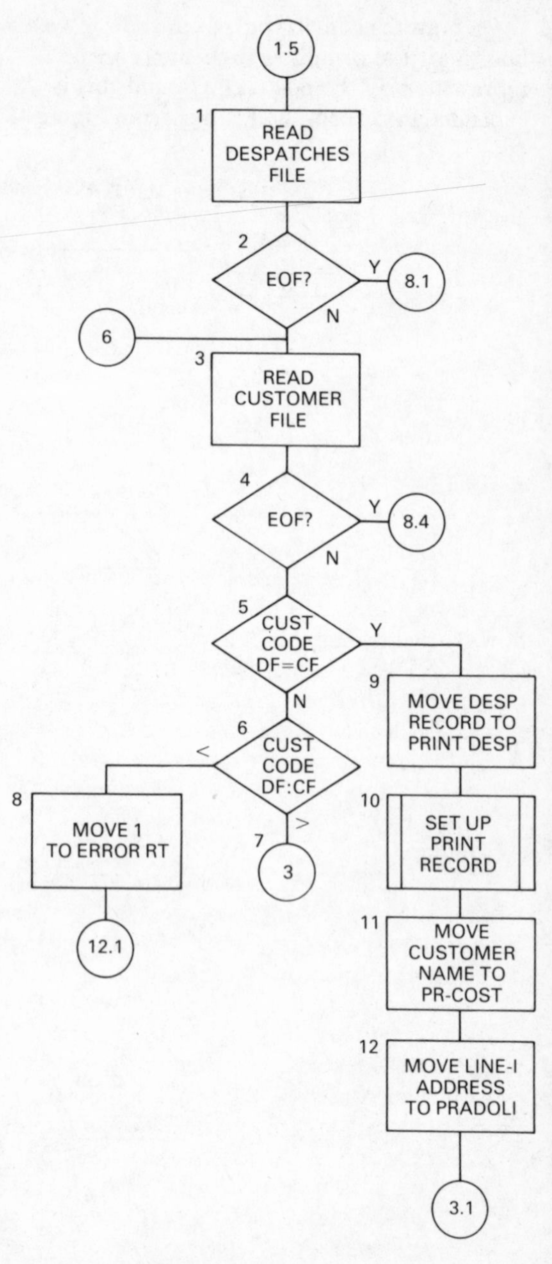

Figure 49 *Example of part of a computer procedure flow chart for COMP 5*

These four flow charts are the main ones that the systems analyst will use. Each of them shows data which is being used in the procedures and the data will be defined separately. They need to be drawn neatly and clearly using standard symbols, if they are to be meaningful tools of communication. The logical flow must be correct and relatively simple decisions should be used; the flow chart provides the analyst with the opportunity to test for repeated processes or missing actions. The logic can be desk checked by passing data through the flow chart and making sure that it works.

Flow charts are a good visual means of communication between the analyst and a user or a programmer; they are reasonably concise and show logical relationships quite well; they are useful for tracing the results of various actions or decisions; and they allow the analyst to analyse and experiment with different ideas on paper. On the other hand, they can easily grow quite complex and when this happens they can be confusing and difficult to use and amend; also it is difficult to trace back from results to actions or decisions. Some of these problems are overcome by the use of decision tables.

Decision tables

Decision tables provide a means of documenting procedures in which several complex decisions have to be made; they are much more concise than flow charts in doing this. They are charts in which all the actions to be performed under all the combinations of conditions are defined. Thus the format of a decision table is as shown in Figure 50. A condition is defined in the condition stub and is one of the factors which must be taken into account in deciding which procedure is to be followed; the collection of all the relevant conditions is the condition stub, and the combinations of conditions which lead to particular procedures is shown in the conditions entry section. An action is defined in the action stub and is one of the steps in the procedure; the collection of all the actions involved in the procedures is the action stub; and the combination of actions which belong to procedures is shown in the action entry section. A rule is a combination of conditions which lead to a particular set of actions. The remarks box is for additional comments.

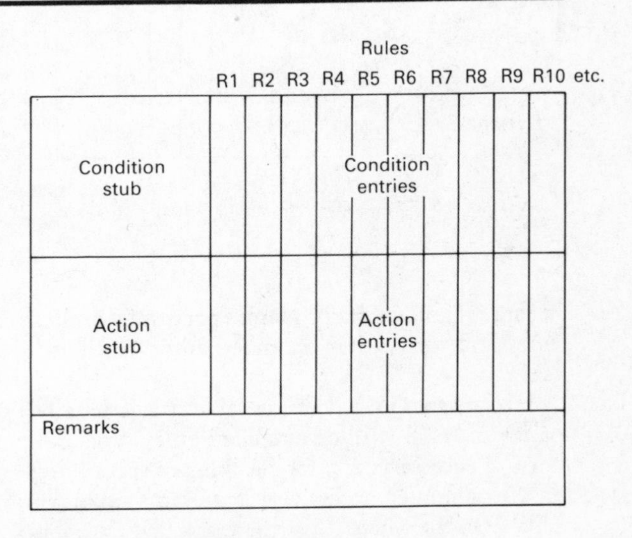

Figure 50 *Decision table format*

There are three types of decision table. A limited entry table is one in which all the conditions are simply expressed – they apply (shown as Y), or they do not apply (shown as N) or it doesn't matter whether they apply or not (shown as –) – and all the actions are simply expressed – they are either performed (shown as X) or not performed (shown as blank). If we draw procedure OTOPR1 (from the system flow chart in Figure 46) as a limited entry decision table it would look like Figure 51.

An extended entry table is one in which some of the conditions and actions can have multiple values, which are shown in the condition or action entries on the right-hand side of the table. If we draw an extended entry decision table for the lower half of the computer procedure flow chart (e.g. symbols 5 to 10 of Figure 49) it would look like Figure 52. This is extended entry because the condition entries are shown as > and < instead of Y or N or –. Some people would call this table a mixed entry table (the third type of decision table)

	R1	R2	R3	R4	R5	R6	R7	R8	R9	R10
If there is an order to deal with	Y	Y	Y	Y	Y	Y	Y	Y	N	N
If it is the 11th, 21st, 31st etc. line	N	Y	Y	Y	N	N	N	Y	–	–
If it is the last line	N	N	Y	Y	Y	Y	N	N	–	–
If the time is 11.30 a.m. or 3.30 p.m.	N	N	N	Y	N	Y	Y	Y	Y	N
Start new computer order		X	X	X				X		
Transcribe line	X	X	X	X	X	X	X	X		
File customer order			X	X	X	X				
Get next orders			X	X	X	X				
Batch all orders completed and send to machine section				X		X	X	X	X	
Go to same table	X	X	X	X	X	X	X	X		
Have a rest									X	X

Figure 51 *A limited entry decision table to define OTOPR1*

If customer code of transaction : customer code of master (symbols 5 and 6)	=	>	<
Set up and print a despatch record (symbols 9 and 10)	X		
Read customer file (symbol 7)		X	
Print error (symbol 8)			X
Read despatches file			X
Go to same table	X	X	X

Figure 52 *An extended entry decision table to define part of COMP 5*

because it follows the principles partly of a limited entry and partly of an extended entry decision table.

In constructing a decision table it is generally easier to begin with an extended entry type and then, if necessary, e.g. for checking or submitting to the computer, to convert it to a limited entry table. On the whole it is true to say that decision tables should not have more than about six conditions or twenty rules because beyond those figures they become very difficult to check.

The steps in constructing a decision table are as follows:

1 Work out the various conditions and write them in the condition stub in any sequence.
2 Work out all the actions and write them in the correct sequence in the action stub.
3 Draw up the rules for all the desired combinations of conditions. (Note that, in a limited entry table, the maximum number of rules will be 2^c where c is the number of conditions; and in an extended entry table the maximum number of rules will be $V_1 \times V_2 \times V_3 \ldots$ where V_1 is the number of values of condition 1, V_2 is the number of values of condition 2 etc.)
4 If all the possible rules have not been shown, add on an ELSE rule as the last rule. This specifies the actions to be taken when any conditions apply which are not specified in the other rules. It will, of course, have no condition entries.
5 Check for ambiguity (for example, where the same combinations of conditions give rise to different actions).
6 Check for redundancy (for example, where the same set of actions is caused by different rules which can therefore be combined by using the hyphen entry).
7 Check the table for accuracy and completeness. The maximum number of rules should be calculated and compared to the actual number. Any rule which has a hyphen or hyphens in it needs to be calculated separately; in this case the number of rules implied by the hyphen is $2 \times H$, where H is the number of hyphens in the rule. Thus in Figure 51 there are four conditions: the maximum number of rules should be $2^4 = 16$; the actual number of rules is 8 without hyphens and 2 with hyphens; the

value of the two with hyphens is $2 \times H = 2 \times 2 = 4$; therefore the actual number of rules (16) is the same as the maximum.

Decision tables are a very useful tool for the systems analyst, especially for defining detailed aspects of clerical or computer procedures. They do not provide an alternative to the flow chart because they serve a different purpose, but in certain circumstances they provide a much more precise and concise way of recording a procedure. An added advantage is that they can often be used for direct entry to the computer, via decision table processors and preprocessors. A processor is usually part of or compatible with a compiler and produces object code directly from the decision table. A preprocessor is a type of compiler that produces source language statements from the decision table to be added later to the source code generated by the programmer. Often these pieces of software place restrictions on the type and size of decision table, but they can also be used to optimize program operation by counting the number of times a particular rule is obeyed and reorganizing the program so that the more frequently used rules come earlier in the program.

The analyst must use his descretion about the size of a decision table, and when the size is too large it should be split down into a number of smaller tables. The tables can be linked together horizontally using GO TO actions or hierarchically using PERFORM actions in the action stub (in the latter case the lower levels of the hierarchy are similar to subroutines of a program).

Structured English

Structured English is the name given to the narrative description of a procedure where the narrative is structured to certain conventions. These conventions are:

1 To define all the steps in a procedure as instructions.
2 To express all the instructions in four different structures:
   (a) Sequential instructions – where each instruction in a sequence is obeyed after the previous one.
   (b) Blocks of sequential instructions – where groups of sequential instructions which are always obeyed together are formed into a block and given a title.
   (c) Decision instructions – where actions which are dependent on conditions are expressed in the form IF . . . THEN . . . ELSE . . . SO . . .
   (d) Loop instructions – where blocks of instructions which are repeated are enclosed in the structure REPEAT . . . UNTIL . . .
3 To write the reserved words of the structuring convention in capitals (i.e. IF, THEN, ELSE, SO, DO, REPEAT, UNTIL).
4 To write the names of blocks of instructions in capitals.
5 To use indentation to clarify the structure.
6 To underline any data (messages or files) which is specified in more detail elsewhere.
7 To use tables to represent CASE structures (i.e. situations where a range of alternatives is mutually exclusive).

An example of structured English defining the procedure COMP5 is given in Figure 53.

A prerequisite to using any form of procedure specification, but especially structured English, is that the logic of the procedure should be clarified and ambiguities removed before the structured English is written. This involves clarifying the conditions and their consequent actions; resolving ambiguities associated with greater than/less than situations (e.g. whether 'greater than 20' includes 20) by using GE and LE (greater/lesser than or equal to) operators; avoiding and/or ambiguities

```
COMP 5
    Read Despatch-file
    IF EOF THEN DO DF–EOF–ROUTINE
    ELSE read Customer-file
    IF EOF THEN DO CF–EOF–ROUTINE
    ELSE IF Despatch–file–customer–code = Customer–file–customer–code
        THEN print despatch–record
    ELSE IF Despatch–file–customer–code > Customer–file–customer–code
        THEN read customer–file
    ELSE (dfcc < cfcc)
        SO print error
           read despatch–file
```

Figure 53 *Example of structured English*

(e.g. take a pencil and a pen or a brush) by using brackets; removing or defining undefined adjectives (e.g. *local* delivery, *most* orders).

Exercises

There follows a description of order processing and stock recording procedures of Mola Ltd, a manufacturer of lawn mowers which holds stocks of spares for a large number of dealers. An order clerk handles all the orders for spares which come directly from dealers; and the same person keeps a record of the stock of each item including amounts on order at the factory and quantities sold to dealers. The ordering procedure is as follows:

'The clerk receives each order and consults his stock records to see whether sufficient stock is available to meet each item on the order. If there is sufficient stock, he adjusts the balance on the stock record and writes the quantity to be despatched on the order next to the quantity ordered.

He then passes the order to the despatch section. Meanwhile he checks the adjusted balance to see if it has fallen below the reorder level; if it has, and there is no outstanding order on the factory, he completes a requisition form for that item.

If there is insufficient stock to meet an order for a particular item, he writes the quantity that is available on the order form and sends it to the despatch department. He then adjusts the balance on the stock record.

If he finds that an outstanding order on the factory is more than one week old, when he is adjusting the balance, he sends an urge note to the factory.

If there is no stock at all to meet an order, the clerk writes zero on the order form next to the quantity ordered, and then phones the appropriate person in the factory to press for a delivery.

Finally, any order which is not completely met is photocopied by the clerk and put into an outstanding orders (dealers) file.'

10.1 Produce a clerical procedure flow chart for this procedure.

10.2 Produce a decision table for this procedure.

10.3 Define this procedure in structured English.

10.4 Compare and contrast your three specifications and suggest which is the best way of documenting the procedure.

Outline solutions to exercises

Chapter 1

1.1 An explanation of the difference between data and information with examples is given on the first pages of Chapter 1.

1.2 The system is the circulation system.
Its subsystems include borrowing books, returning books, reserving books etc.
The interfaces between these subsystems take the form of lists of books borrowed by whom.
The inputs are borrower identifier, book identifier, and date.
The outputs are lists of borrowers, lists of overdue books, lists of reserved books etc.
The control function is triggered by time which determines when books should be returned.
The system adapts to its environment by making sure that the accession subsystem acquires appropriate books for borrowers, that careful control is exercised over overdue returns, that users of the library are carefully vetted, etc.

1.3 (a) A stock control application would offer the benefits of a reduction in stock levels, an improvement in customer service, a balanced stock range in relation to demand and an efficient replenishment procedure. This would be achieved by centralization of stock records and timely updating of the records with details of all issues and withdrawals.
(b) A personnel records system would aim to provide more comprehensive, integrated and up-to-date information about employees to assist in manpower planning and allocation. It would achieve this through the computerization of personnel records and the early updating of these records with details of leavers, starters, and changes in the circumstances of existing employees. It would probably be linked to the payroll system.
(c) A traffic control system would offer the benefits of better road utilization, reduced traffic congestion, fewer accidents and closer supervision of traffic flow. It would achieve this by counting/measuring traffic flow and regulating the control points (i.e. traffic lights, speed controls, lane controls etc.) to smooth the flow.

1.4 (a) The information needs of a careers adviser are mainly concerned with rapid access to a large amount of information about jobs/courses, vacancies and applicants. (Each of these topics should be expanded in greater detail.)
(b) The computer might be used:
To maintain large databases of up-to-date information on jobs, courses, vacancies and applicants.
To allow applicants to browse through jobs and courses information.
To carry out a certain amount of matching of applicants to courses/jobs.
To gather information about an applicant through an interactive dialogue.

Chapter 2

2.1 Systems analyst → top managers: overall system proposals for evaluation; progress reports.
Top managers → systems analyst: terms of reference for projects; feedback on proposals/progress reports.
Systems analyst → user managers: reports on requirements, system proposals for approval.
User manager → systems analyst: information about requirements for new system in light of responsibilities, plans,

problems, decision areas etc.
Systems analyst → auditor: plans for and details of controls/security measures in new systems.
Auditor → systems analyst: weaknesses in systems, audit requirements.
Systems analyst → external bodies: plans for/details of new system which affects external bodies.
External bodies → systems analyst: rules, regulations and requirements of trade associations, government departments etc.
Systems analyst → clerks/shop-floor operatives: ideas about possible new systems.
Clerks/shop-floor operatives → systems analyst: information about current working practices/problems.
Systems analyst → other systems analysts (and vice versa): ideas/specifications/ findings about systems which may affect other systems/subsystems.
Systems analyst → programmers: program specifications, amendments, clarification.
Programmer → systems analyst: advice on technical aspects of design, queries, tested and working programs.
Systems analyst → data control and data preparation: instructions for control and input procedures.
Data control and data preparation → systems analyst: problems, constraints, advice on control and input procedures.
Systems analyst → computer operators: instructions for operating procedures (especially error and recovery procedures).
Computer operators → systems analyst: problems, constraints and advice on operating procedures.

2.2 The sales assistants may have this view because:
It is a poor system (no feedback given to store about state of order).
They have been poorly trained (they don't seem to know how the system works).
They rather resent, and are hostile towards, the computer system (because they don't understand it, perhaps, or because it was imposed on them).
They are giving an excuse for an error/omission of the system.
The manager should:
Discover the actual reason.
Look at the system, the training, communication within the store to see if they can be improved.
Examine the possibility of running a computer appreciation course for staff.
Encourage staff to get involved in/make suggestions about improvements to the system.

2.3 The three advertisements are intended to recruit one person who will have the particular personality, skills and knowledge that are felt to be appropriate. Each advert should mention all of these but emphasize different aspects in relation to the particular background of the people whom the advertisement is aimed to attract. The general requirements might be for someone with an extrovert personality who gets on well with people but is quite analytical; with knowledge of both computing techniques and production applications; and with skills in the whole range of systems analysis techniques from interviewing to program design.

The advertisement aimed at the programmers would probably emphasize the need for an analytical approach, computing expertise and design skills. The advertisement aimed at user staff would highlight the personality required and knowledge of production systems, whilst offering training in systems analysis skills. The advertisement aimed at polytechnic graduates would emphasize the need for their knowledge and skills to be in the area of commercial systems analysis with a good understanding of business applications.

2.4 Depends on the reader's own experience of a data processing department.

Chapter 3

3.1 Decay in a patients' register system might be due to:

Changes in the purpose of the system.
Information becoming inaccurate and out of date.
Changes in public attitude to privacy (and security of confidential information).
Increased volumes of data.
Obsolete technology (hardware or software).
Changes in volatility of files.
Changes in medical knowledge.
Changes in staffing.

3.2 The social feasibility of a patients' record system would need to be looked at from the point of view of the users (medical, administrative, clerical), the patients whose details are recorded, and the community at large. Some of the issues which would need investigation are:

Impact on jobs (location, content, satisfaction).
Need for education/training.
Union/staff association/BMA views.
Mechanisms for user involvement/consultation.
Privacy of medical information.
Security measures.
Procedural impact on patients.
Costs (especially administrative versus medical budgets).

3.3 More time might be spent on feasibility because the computerization of personal details (especially medical details) is very sensitive. Also the system would be large and complex but would need to offer fast access to up-to-date information – the costs and the technical problems of this are considerable. Implementation would be more time consuming because of the very large number of users who would need to be trained to use the system.

Chapter 4

4.1 A large number of people could be seen to have an interest in a computerized pupil record, including:

Pupils
Parents
Teachers
Head teachers
Administrators of the local education authority
Careers Advisors
Educational psychologists
Education welfare officers
School health service

All of them could reasonably participate in a feasibility study, as suppliers and users of the stored information. All except the administrators could potentially be suppliers of personal information whose confidentiality is important; they would want to contribute to decisions about what information should be collected, for what purpose, how long it would be retained, how its accuracy could be assured and who would have access to it. The professional staff would be concerned about the feasibility of coding subjective judgements about academic, psychological, medical or social aspects of a child's progress. The critical issue revolves around whether any significant benefits could be achieved for decision-making about individual pupils whilst retaining confidentiality of information; up-to-date information from a central source would clearly be valuable but there are severe problems in collecting and maintaining the accuracy of such information. The administrators would benefit from such a system because more accurate statistics would be available for planning such things as school size, location, transport, meals, etc.

4.2 The alternatives are:

Development

1 In-house
2 Contracted to consultancy
3 Co-operative
4 Modification of package.

Running

5 In-house
6 Bureau

{ batch versus demand, on-line versus off-line, local versus remote

The most common combination is 1 with 5; a decision to adopt another approach would be based on financial evaluation (2, 3, 4 might be cheaper than 1; 6 might be cheaper than 5). Factors which would favour the other approaches might be:

2 In-house staff too busy, first application, time-scale very short, specialized expertise needed.
3 Co-operation between regions of a large decentralized organization (e.g. health authority regions, electricity boards, national outlets of multinational organization) or between small firms in a trade association.
4 Package available which has option facilities or can easily be modified and which is fairly close to the organization's needs; or which meets very generalized data processing needs (e.g. data management, teleprocessing).
6 In-house machine overloaded, specialized application, first application.

4.3 The solution to this question depends on the applications identified in Exercise 1.3. A list of benefits usually looked for might be:

Quantifiable savings:

1 Reduction in staff, accommodation, supplies, equipment
2 Handling of higher volumes with current staff
3 Avoidance of peaking problems
4 Faster processing cycle – better cash flow
5 Reduced stocks, WIP, warehouse space
6 Increased sales
7 More control over expenditure.

Less quantifiable gains:

8 Better information → better decisions
9 Better planning
10 Improved morale/image
11 Better communication in organization
12 More disciplined procedures
13 Application of management techniques.

The benefits of a stock control system (Exercise 1.3) would be mainly quantifiable (1, 5, 6, 7, 12); the benefits of a personnel records system would be much less quantifiable (8, 9, 11, 13); and a traffic control system would be in between (2, 4, 8, 9, 13).

4.4

Costs outlay £150,000		*NPV at discount rate* 10%	12%	14%
Year	*Income*			
1	£13,000	£ 11,817	11,609	11,401
2	£35,000	£ 28,910	27,895	26,915
3	£35,000	£ 26,285	24,920	23,625
4	£70,000	£ 47,810	44,520	41,440
5	£75,000	£ 46,575	42,525	38,925
Profit	£78,000 *NPV*	£ 11,397	1,469	–7,694

The recommendation to management would be that the project should go ahead only if the rate of inflation is certain to be no more than 12 per cent p.a. during the life of the project.

The reduced development cost would show the following result:

Cost outlay	£140,000	*NPV* *at 10%*	*at 12%*	*at 14%*
Profit	£ 63,000	9580	–140	–9095

In this case the project should only go ahead if the rate of inflation is certain to be less than 12 per cent p.a. during the life of the project.

Both recommendations would need, of course, to be compared to other investment opportunities available to the organization.

Chapter 5

5.1 The background information might include:
The economic and political situation in the local authority (constraints and possibilities).
The short- and long-term environmental plans (from the planning department).
Past studies on traffic control (internally and externally).
Current policies *re* bus lanes, parking, ring roads, pedestrianization, one-way traffic systems etc.
Current procedures/methods of traffic control in different local authorities.

5.2 Extra information would be gathered by:
(a) Observation of the roads under consideration and the way in which they are used by private vehicles, pedestrians and public transport (especially the flow, any congestion and parking situations).
(b) Searching existing records such as maps, accident reports and statistics, maintenance work reports, minutes of meetings etc.; it would also be appropriate to use equipment to collect records of road utilization (e.g. mechanical counters or VTR cameras).
(c) Sending questionnaires to a sample of road users (e.g. drivers, cyclists, pedestrians) about their journey patterns.

5.3 Interviews would be conducted with such people as planners (about possible changes to the road system), engineers (about road furniture, construction and maintenance), police and other services (about accidents and problems of flow/access), and a small sample of road users (about their view of traffic flow and how it could be improved).

Checklist of points for interview with police officer responsible for accident statistics:
Introduction
Purpose of investigation
Overall departmental structure
Responsibilities, duties
Data collection procedures
Forms used
Analysis procedures
Report formats and contents
Information specific to area to be controlled
Meaning of this information
Possible improvements which are needed
Summary
Thanks

5.4

Erewhon County Council
Road traffic survey

Please complete this brief questionnaire to help in improving traffic conditions in Erewhon.
Tick the appropriate box in answer to each question.

1 How often do you travel into Erewhon?
Daily ☐ Weekly ☐ Monthly ☐

2 How do you mainly travel into Erewhon?
By car ☐ On foot ☐ By cycle ☐
By bus ☐ By train ☐

3 How long is your journey?
Less than 15 min ☐ 15–30 min ☐
30–45 min ☐ more than 45 min ☐

4 Which route do you mainly use?
M49 ☐ A43 ☐ A416 ☐ other ☐

etc.

Chapter 6

6.1 Input data:
- Exam no.
- Centre no.
- Marker/examiner no.
- Mark details
  - Candidate no.
  - Mark

GCE exam marks

Examination no. ☐☐☐☐☐☐ Examination name ________

Centre no. ☐☐☐☐☐☐ Centre name ________

Marker no. ☐☐☐☐☐☐ Marker name ________

Candidate no.	Mark	Candidate no.	Mark	etc.

6.2 Ideal case for preprinted document to collect marks – printed by computer – with marks (only) keyed in at VDU into formatted screen. In this case all data except marks checked visually; marks checked by range check.

6.3 Files:
- Exam:
  - Exam no.
  - Exam name
- Markers:
  - Examiner no.
  - Grade
  - Exam details
    - Exam no.
    - Fee rate
    - No. of scripts
- Exam entries/candidates:
  - Candidate no.
  - Candidate name and address
  - School no.
  - Centre no.
  - Exam details
    - Exam entered no.
    - Predicted result
    - Actual result
- Schools:
  - School no.
  - School name and address
  - Exam details
    - Exam no.
    - Passes by grade by year

All files should be stored indexed sequentially (to allow both sequential and direct processing) on disk.

6.4 (a)

GCE RESULTS – ANALYSIS OF MARKING

EXAM NO. XXXXXX EXAM NAME XXXXXXXXXXXXXXXXXXXXXXXXXX

MARKER NO.	MARK RANGE							
	< 10	11–20	21–30	31–40	41–50	51–60	61–70	71–80 etc.
XXXXX	XX	XX	XX	XX	XX	XX	XX	XX

MEAN

(b)

STUDENT RESULT REPORT

EXAM NO.	STUDENT NO.	STUDENT NAME	RESULT	EXPECTED RESULT
XXXXXX	XXXXX	XXXXXXX	XX	XX

GCE RESULTS BY SCHOOL

SCHOOL NO. XXXXXXXX SCHOOL NAME ______________

EXAM NO.	GRADE A					THIS YEAR	GRADE B			
	Yr 1	Yr 2	Yr 3	Yr 4	Yr 5		Yr 1	Yr 3	Yr 3	etc.

The students results would come first and the summary by examination second.

(c)

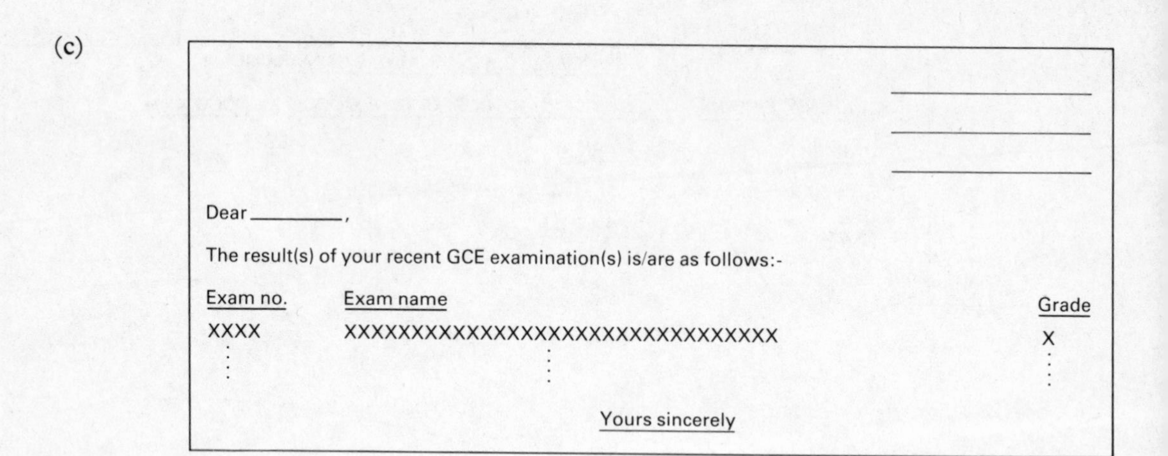

Dear ________,

The result(s) of your recent GCE examination(s) is/are as follows:-

Exam no.	Exam name	Grade
XXXX	XXXXXXXXXXXXXXXXXXXXXXXXXXXXXXXX	X
⋮	⋮	⋮

Yours sincerely

(d)

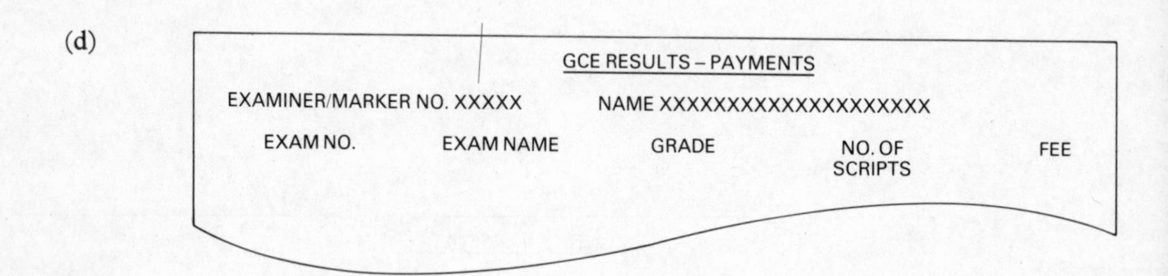

GCE RESULTS – PAYMENTS

EXAMINER/MARKER NO. XXXXX NAME XXXXXXXXXXXXXXXXXXXX

EXAM NO.	EXAM NAME	GRADE	NO. OF SCRIPTS	FEE

6.5

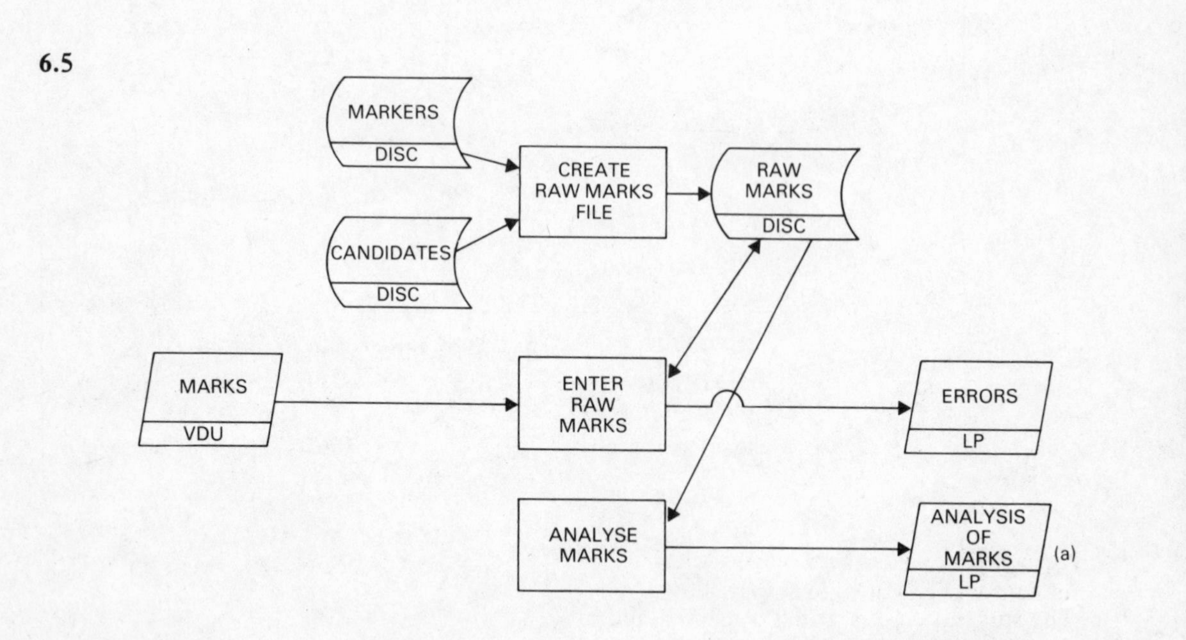

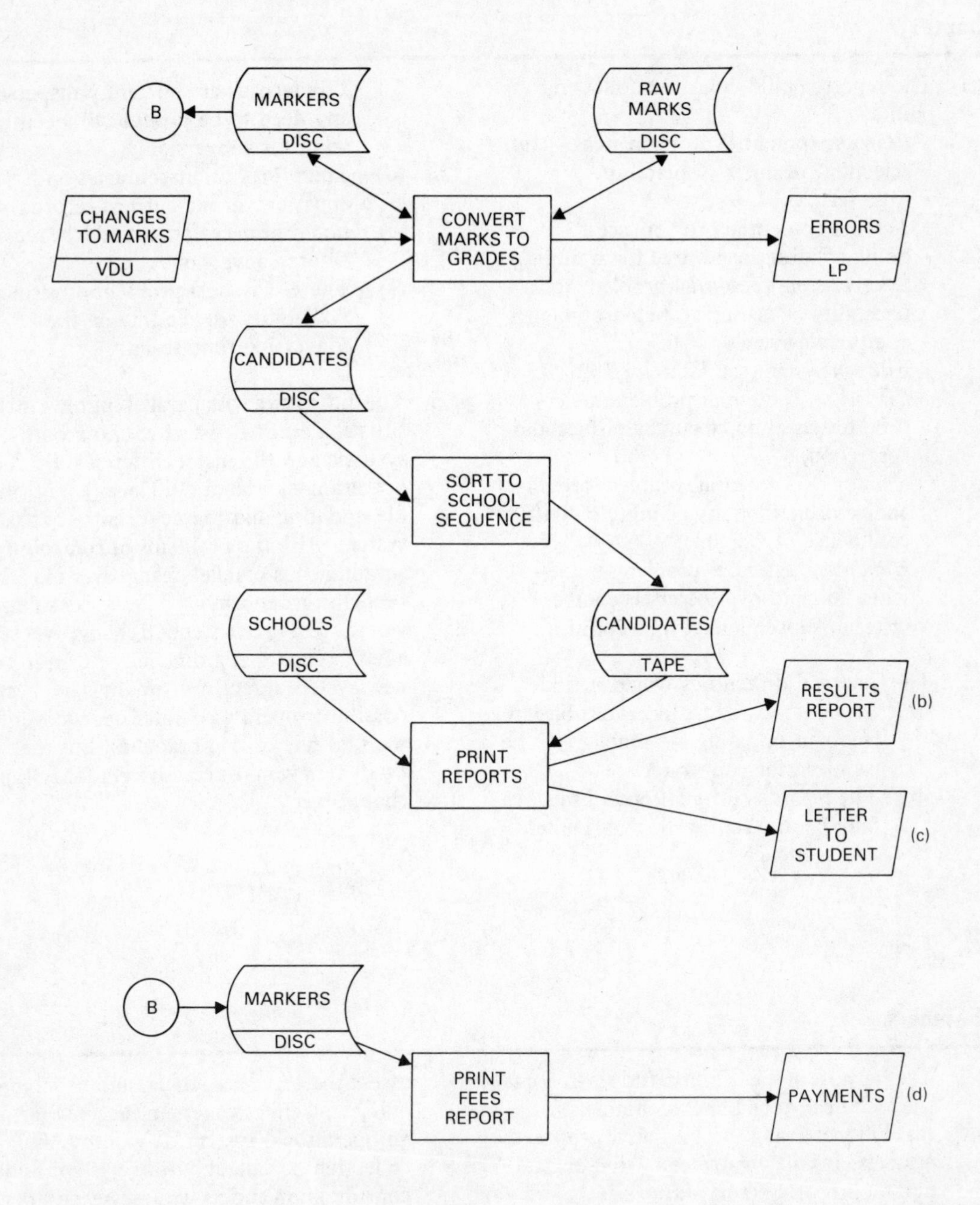

The control totals at each stage could be number of candidates, total marks.

Chapter 7

7.1 The report should cover the following points:

Planning (potential problem areas – staff selection, resource availability, time-scale).

Training (potential problem areas – users doing training, users free for training).

System testing (potential problem areas – feasibility of testing real environment, comprehensiveness of tests).

File conversion (see Exercise 7.2).

File set-up (potential problem areas – time for checking accuracy of files and correction).

Changeover (potential problem areas – choice of method, availability of staff, confidence of users).

Handover (potential problem areas – when to hand over, user acceptance criteria, subsequent responsibilities).

7.2 The potential difficulties would include:

(a) Account records are live and subject to frequent updating – cannot easily be removed for conversion.

(b) The records at the different branches may be different in format, content and sequence.

(c) The records are currently dispersed and need to be brought to a central point for conversion.

(d) There may be inaccuracies and omissions in the current records.

(e) It may be necessary to code the records before conversion.

(f) There may be a problem of resource availability, depending on the conversion time-scale.

7.3 The difficulties with parallel running in this situation are the cost of running both systems and the major difference in operation between the old locally controlled file updating and the new centrally based system. Also the problems of controlling a simultaneous parallel changeover in eighty branches are enormous. The better solution would seem to be a stepped changeover with a few branches at a time moving on to the new system directly. It might have been possible to operate a parallel changeover for the first one or two branches, but the time-scale would necessitate a fairly rapid changeover

Chapter 8

8.1 The purpose of a feasibility study is to assess the technical, social and economic feasibility of a range of alternative proposed systems. It will involve collection of data about current systems and user requirements; users will participate in identifying their needs; and a development plan will be produced. The feasibility study is one-off per system and is concerned with forecasting benefits.

The purpose of a system review is to assess the technical, social and economic efficiency, effectiveness and acceptability of an operational system. It will involve collection of data about the system under consideration and its weaknesses; users will contribute to the identification of potential changes. The system review should be conducted periodically and its emphasis is on measuring whether forecast benefits have been achieved.

8.2 **The form might look like:**

Amendment specification form

Originator | Author

Purpose of amendment

Specification of amendment

Verified by | Priority

Authorized by

Action

8.3 Points for charging:

Increases user involvement/commitment
Reduces the number of 'frivolous' applications
Improves planning and control
Facilitates more quantitative evaluation
Encourages computer staff to be more realistic in estimates
Allows computer department to operate as profit centre with own budget

Points against charging:

Increases reluctance of users to take risks
Militates against integrated developments
Users may choose to go outside for the service

Chapter 9

9.1 The type of information which needs to be gathered would include:
(a) Cash register with cassette storage: cash register would look and work like normal cash register but would have additional keys for product code to be entered for each transaction. The product code and amount would then be recorded on cassette which can be sent at appropriate intervals to the computer centre. The cost would be about twice the price of a normal cash register and of course every shop would need at least one new cash register – so it would be very expensive. On the other hand, staff are used

to using cash registers – and also the data would be collected as a by-product of entering cash transactions.
(b) A Kimball tag is a small card (can be produced in multiparts) into which up to forty characters of data can be punched as small round holes (like punched paper tape but smaller). The tag can be attached to the goods and then removed at the time of sale – thus the Kimball tags can be sent by first post to the computer centre to provide early and accurate updating of files. The Kimball tag system is very simple and easy to operate, and is relatively cheap.
(c) A hand-held scanner for reading bar-coded labels usually has a cassette recorder attached to it for recording the label data which has been read. The cassette would be sent as and when required to the computer centre. The label would have a bar-marked product code to capture data about stock movements. The system is relatively cheap and very easy to use. The scanner can also be used for regular stock-taking.

9.2 The report should consist of:
(a) An introduction (stating the purpose of the investigation and how conducted);
(b) The detailed findings about the equipment;
(c) An evaluation of the different methods *vis-à-vis* other requirements;
(d) A recommendation.
I would suggest that the light pen approach is best because it is relatively cheap and easy to use and it has less security/control problems than Kimball tags.

9.3 Depends on the method recommended but the talk should cover:
Why? (purpose of the change)
What? (what the change is)
How? (how the system will work)
When? (when to be implemented)
Who? (effect on staff)
Where? (any location changes)

9.4 Depends on the method recommended, but the visuals should be clearly drawn, very specific and not too wordy.

Chapter 10

10.1 A possible procedure flow chart is as shown opposite.

10.2

	R1	*R2*	*R3*	*R4*	*R5*	*R6*	*R7*	*R8*	*R9*	*R10*
Stock available > 0?	Y	Y	Y	Y	Y	Y	Y	N	N	N
Stock available > quantity ordered?	Y	Y	Y	Y	N	N	N	N	N	N
Stock available – quant. ord. > reorder level?	Y	N	N	N	N	N	N	N	N	N
Outstanding order on factory?	–	N	Y	Y	N	Y	Y	N	Y	Y
Outstanding order on factory > 1 week?	–	N	N	Y	N	N	Y	N	N	Y
Adjust stock balance	X	X	X	X	X	X	X			
Enter quantity despatched	X	X	X	X	X	X	X			
Enter zero								X	X	X
Complete requisition		X			X			X		
Send urge note				X			X			X
Phone to press for delivery								X	X	X
Photocopy order and place in O/O file					X	X	X	X	X	X
Send order to despatch	X	X	X	X	X	X	X	X	X	X

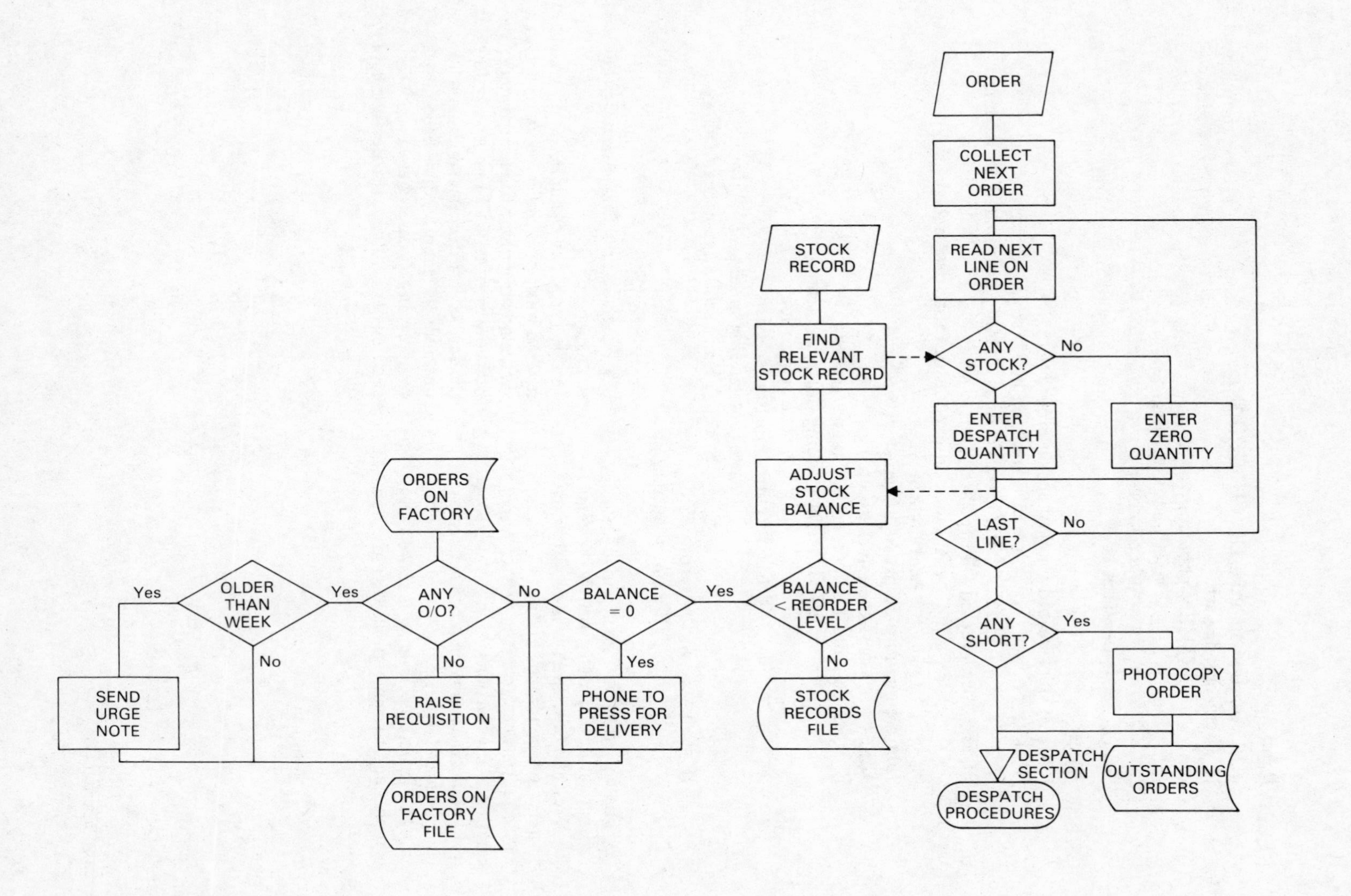
ORDER
COLLECT NEXT ORDER
READ NEXT LINE ON ORDER
STOCK RECORD
FIND RELEVANT STOCK RECORD
ANY STOCK?
No
ENTER DESPATCH QUANTITY
ENTER ZERO QUANTITY
ADJUST STOCK BALANCE
LAST LINE?
No
ORDERS ON FACTORY
Yes
OLDER THAN WEEK
Yes
ANY O/O?
No
BALANCE = 0
Yes
BALANCE < REORDER LEVEL
ANY SHORT?
Yes
No
No
Yes
No
SEND URGE NOTE
RAISE REQUISITION
PHONE TO PRESS FOR DELIVERY
STOCK RECORDS FILE
PHOTOCOPY ORDER
DESPATCH SECTION
OUTSTANDING ORDERS
DESPATCH PROCEDURES
ORDERS ON FACTORY FILE

10.3 RECEIVE-ORDER

```
        FOR each line on order DO CHECK-LINE
        IF Photocopy-flag set
                 THEN photocopy order
                      reset photocopy-flag
                      file photocopy
        Send order to despatch

CHECK-LINE
        Retrieve stock-record for product on order-line
        IF stock-ordered (order-line) LE stock-on-hand (stock-record)
                 THEN write stock-ordered as quantity despatched (order-line)
        ELSE (stock-ordered GT stock-on-hand)
                      write stock-on-hand as quantity despatched (order-line)
                      set photocopy-flag
        Subtract quantity-despatched from stock-on-hand
        Adjust stock-on-hand (stock-record)
        IF stock-on-hand LT reorder-level (stock-record)
                 THEN IF no outstanding-order on factory
                           THEN send requisition
                      ELSE (outstanding-order exists)
                           IF date (outstanding-order) GT one week
                                    THEN send urge-note to factory
        IF stock-on-hand = zero
                 THEN phone factory
```

10.4 (a) The flow chart is a good way of experimenting pictorially with the problem.
(b) With the flow chart it is easy to trace actions from conditions but not vice versa; with the decision table it is easy to trace conditions from actions.
(c) The decision table can be checked for completeness.
(d) The structured English is more compact and less ambiguous than the flow chart.
(e) The structured English is relatively easy to change.
In this particular example the preferred method must be a choice between decision table and structured English. In most procedures structured English would carry the day because decision tables soon become unwieldy as the procedure becomes complex; in this case where a small series of decisions is involved, the decision table is quite useful.

Index